UNMASKING EUROPA

UNMASKING EUROPA

THE SEARCH FOR LIFE ON JUPITER'S OCEAN MOON

Richard Greenberg

COPERNICUS BOOKS
An Imprint of Springer Science+Business Media

In Association with
PRAXIS PUBLISHING LTD

ISBN 978-0-387-47936-1 e-ISBN 978-0-387-09676-6

Published in the United States by Copernicus Books,
an imprint of Springer Science+Business Media.

Copernicus Books
Springer Science+Business Media
233 Spring Street
New York, NY 10013
www.springer.com

Library of Congress Control Number: 2008929493

Cover Illustration: NASA Jet Propulsion Laboratory (NASA-JPL)

Manufactured in the United States of America.
Printed on acid-free paper.

9 8 7 6 5 4 3 2 1

Contents

Preface

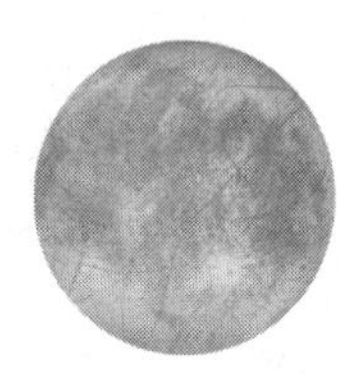

My book *Europa, the Ocean Moon: Search for an Alien Biosphere* was published for the academic science market. Some individuals beholden to the US planetary exploration establishment really hated it, because I intertwined a story that ordinarily doesn't go into an academic science book. They didn't like what I had to say about how science is done in the context of a major big-ticket space mission, and about how the human interactions can skew the canonical scientific results. One reviewer wrote that those parts moved him to throw the book to the floor. I wasn't bothered too much because (a) the reviewer was someone I had portrayed in the book so he certainly had a conflict of interest and (b) the review resulted in a giant sales surge (relatively speaking—that was a academic book). Obviously, I had touched a nerve.

On the other hand, I got some great reviews and a lot of fan mail, especially from Europeans, aerospace engineers, amateur space enthusiasts, students, biologists and various combinations thereof—more generally, people willing and able to wade through the technical detail, but without a reverential commitment to the US planetary-exploration establishment.

I felt from the start that the topics I covered could have much broader appeal: The scientific subject is the cold, wet place where we are most likely to find extraterrestrial life; The political aspects mirror the broader 21st century world, where spin trumps substance, anecdotes trump systematic evidence, and fabrications are given more credence than carefully developed arguments; And the human story is about the gritty creative process of doing science. At the frontier, scientists face the same agonies, insecurities, conflicts, and uncertainties as their counterparts in any of the other creative arts. When you are pushing the boundaries, you don't know what you are doing or how it will come out, but you have to put yourself on the line. So any true story about doing science is about intense feelings—the pain and elation of creativity and discovery.

Interesting and controversial as the human story was, it remained embedded in all the arcane detail of the academic book, with mathematical equations and a price that ensured a small specialized readership. Now I have stripped away much of the academic detail so the story can be reached by a wider audience. I hope this book will give you a good idea about what Europa is like and, just as important, the sometimes bumpy and circuitous and even blindly lucky path traveled by scientists as they make their way toward understanding.

The scientific part of the story is about how we developed the evidence that Europa's surface lies on only a thin permeable ice crust, with ongoing dynamic connections to a liquid ocean just below. This view of Europa, with its implications for extraterrestrial life, is controversial, but as I describe in the book, the controversy arose less from reasoned deliberation than from political ambition and manipulation.

Our understanding of Europa comes largely from the results of NASA's Galileo mission to Jupiter and its moons. This book describes that project from a particular perspective—an insider who rarely bought the party line. Along the way I criticize aspects of the Galileo project, but only where caution and conservatism were extended inappropriately to scientific analysis, discouraging creativity and risk-taking just where they were most needed. But without the care and dedication of the engineers and administrators of the Galileo project the scientific dimension of this story could never have been told. Those remarkable people got the robot to Jupiter and they made it work, even as bad luck kept throwing challenges in the way. The discoveries described in this book could not have been made without them.

The picture of Europa developed here is due to the insights and hard work of several remarkable scientists who have worked in my research group. Many of the key discoveries were made by my post-doctoral associate Paul Geissler and my students Greg Hoppa and Randy Tufts. Michael Benson's 2003 *New Yorker* article on the end of the Galileo mission highlighted Paul, Randy, and Greg's discoveries. This book puts them in a more complete context. Other students and associates in my group who contributed in important ways to our Europa work were Dave O'Brien, Alyssa Sarid, Terry Hurford, Jeannie Riley, Martha Leake, Dan Durda, Gwen Bart, and Susan Arthofer. I am deeply grateful for all that they taught me and for the pleasure and comfort of their friendship. Their collective work makes up much of the scientific narrative of this book.

I had the opportunity to connect with such great people because I work at one of the premier centers of research in planetary science, the University of Arizona's Lunar and Planetary Laboratory, with our academic arm the Department of Planetary Sciences. The leadership of Mike Drake and his predecessor Gene Levy have nutured an environment where creativity, intellectualism, discovery, and innovation flourish. The faculty and staff of LPL make this a wonderful place to work and many of them (too many to name them all) contributed to the Europa story in a wide variety of ways. Selina Johnson managed the affairs of our research group. Jay Melosh helped me and my students understand various crucial geophysical issues. LPL's Planetary Image Research Laboratory (PIRL), under the direction of Alfred McEwen, facilitated our immediate access to Galileo data as they arrived on Earth, and produced a range of image products that were essential for our investigations. Most of the images in this book were processed at PIRL, including important mosaics and color versions produced by students Cynthia Phillips and Moses Milazzo. Joe Plassman, Chris Schaller, and Zibi Turtle of the PIRL group helped in many ways.

The research by my group described in this book was supported by NASA through the Galileo project, the Planetary Geology and Geophysics program, and the Jupiter Data Analysis program and by NSF through its Life in Extreme Environments program. Several chapters are based on material that we published in *Icarus*, and the section on planetary protection is based on our article in *American Scientist*.

This book came to be thanks to the confidence and encouragement of Clive Horwood of Praxis Books and Paul Manning of Springer. I am also grateful to Harry Blom of Springer and Copernicus Books for his unwavering commitment to the project. I think Harry was afraid to tell me what Paul Farrell said about the manuscript at first, but eventually Paul came to understand what I was trying to do, and signed on as editor. Almost every suggestion by Paul made the book better, and I am very grateful to him. Philippe Blondel of the University of Bath edited the academic book for Praxis, and many of his improvements have carried over to this book as well. Olga Gómez Sichi read an early manuscript for Praxis and provided some important corrections and suggestions. Henry Kandel of Saint Ann's School in Brooklyn, New York, provided early insightful comments that convinced Paul Farrell to get on board. My dad Elliot Greenberg read it too; he said he liked it. Greg Hoppa counseled me throughout the process of writing both books, and helped me see things from a more optimistic perspective. Dickinson College in Carlisle, Pennsylvania, provided the facilities for writing this book as well as the earlier academic book, during two extended visits. I am especially grateful to the Dickinson Physics and Astronomy faculty, my hosts Priscilla and Ken Laws, and department chairs Hans Pfister and Robert Boyle.

In academia your closest research associates tend to move on every few years. That's the way it is supposed to be. My Europa gang has dispersed now, but they are mostly out doing interesting things. Paul Geissler is concentrating on Mars studies now. Greg Hoppa, after reaching a point of deep personal satisfaction with his seminal contributions to the understanding of Europa, realized that his prospects in academia were poor, and needing to support his growing family, he took a job in the aerospace industry, unrelated to planetary science. The field lost one of its most capable and creative young scientists. Randy too was contemplating the next stage in his career when he learned he had a bone-marrow disorder. Before he died in 2002, in addition to all his other accomplishments, he had helped revolutionize our thinking about the ocean moon.

Future investigations and exploration will test our work, but nothing will top the thrill and satisfaction we all shared in seeing the pieces of the puzzle come together as we came to understand and to love Europa.

A note to the reader: Web sites for images

This book is about looking at images, and it is full of them. In addition, the entire treasure trove of spacecraft images of Europa is available on-line. The following are some useful sources.

NASA's archive of spacecraft mission data is stored in the Planetary Data System. The following site contains Europa imaging data. These are the scientific images, relatively unprocessed, and presented without interpretation. Select Europa, and one of the orbits, and then take your choice from the thumbnail images of those you want to download at full resolution. Note that on the web site the imaging camera is called SSI, and the Near-Infrared Mapping Spectrometer is NIMS:

http://pds-imaging.jpl.nasa.gov/cgi-bin/Nav/GLL_search.pl

The Galileo web site includes numerous miscellaneous image products, but beware of graphics, captions, and interpretations that may be out of date or misleading:

http://www2.jpl.nasa.gov/galileo/images/europa/eurimages.html

JPL's Planetary Photojournal includes many press release products. Labels starting with "PIA" are definitely public information versions, not science products. As with the Galileo site, graphics, captions, and interpretations that may be out of date or misleading. But viewed with that caveat in mind, these can be very interesting images.

http://photojournal.jpl.nasa.gov/target/Europa

Water World

"Europa! I found it."
"That's Eureka, Maynard."

—*Maynard G. Krebs and Dobie Gillis*

Brisk tidal water sweeps over creatures clawed into the ice, bearing a fleet of jellyfish and other floaters to the source of their nourishment. As the water reaches the limits of its flow, it picks up oxygen from the pores of the ice, oxygen formed by the breakdown of frozen H_2O and by tiny plants that breath it out as they extract energy from the sun. The floating creatures absorb the oxygen and graze on the plants for a few hours.

The water cools quickly, but before more than a thin layer can freeze, the ebbing tide drags the animals deep down through cracks in the ice to the warmer ocean below. Most of the creatures survive the trip, but some become frozen to the walls of the water channels, and others are grabbed and eaten by anchored creatures waiting for them to drift past. The daily cycle goes on, with plants, herbivores, and carnivores playing out their roles.

Tides support rich and varied life all around our planet. Tide pools are rich in plants and animals that exploit the boundary between sea and land, moving with the tidal flow, or anchoring themselves (as do mussels or seaweed, for example) to be fed and flushed by the ebb and flow. Coastal marshes are fecund with tide-supported life. Creatures from land, sea, and air gather at these tidal margins to feed on the riches. Even in the Arctic, under conditions that seem harsh to our species, rich and diverse ecologies live with the rhythm of the tides.

But the scene described above might occupy a setting far more exotic: the icy crust of Jupiter's moon Europa. Europa is about as big as our own Moon, but it is an ocean world. Compared with our Moon it is over five times farther from the Sun and its planet, Jupiter, is about 300 times the mass of the Earth. The great distance from the Sun means that Europa formed with less rock and plenty of water, and the enormous mass of Jupiter generates active tides on Europa. Tides stress and crack the ice, and heat from tidal friction is responsible for the existence of the ocean itself.

An active crack in the ice crust that links the ocean to the surface might support the daily cycle of life described above, as tides squeeze liquid water up and down between the ocean and the surface. Generations come and go as the daily cycle repeats for thousands of years. But longer than that, the tidal rhythm will change, so that any particular crack must eventually freeze shut. Life will go on only if it can adapt to change. Some organisms could make their way to another more recently opened crack. Others, frozen into place, must be able to hibernate as some bacteria do on Earth. Their wait would be rewarded by new cracking or by a melting event, either of which will bring back the warm oceanic water.

Of course, nobody on Earth knows whether life exists on Europa or ever did. Conventional speculation about life beyond our own planet has assumed that it must lie close enough to the Sun for warmth, but not so close that there is no water. Those criteria make Mars a good candidate. Astrophysicists contemplating the hundreds of new planetary systems recently discovered around other stars refer to the "habitable zone" as the narrow range of distances from a star where Earth-like planets might exist. By that criterion, Europa lies way outside the habitable zone of our solar system.

Nevertheless, thanks to tides, physical conditions on Europa appear to provide settings that could support life. There are likely a variety of niches that might be home to complex ecosystems. And while environmental changes over various scales of time and place would challenge life, they might provide opportunities for diversification and evolution.

This picture has emerged as we have come to know Europa as a planet. Although Europa is formally a moon, planetary scientists often lapse into calling a moon a "planet" once its detailed character is revealed to us, and especially if that character and the processes involved appear to be planet-like.

Exploration of Europa began with Galileo's early telescopic discoveries, and Earth-based observations continued for hundreds of years as instruments continually improved. Since the start of the space age, two spacecraft have given us close-up views. Voyager 2 gave us our first good look at Europa's surface during its quick fly-by in July 1979. (Its twin, Voyager 1, also flew past Jupiter but did not get close to Europa.) The follow-up mission, originally named the Jupiter Orbiter with Probe, was planned to orbit Jupiter during the 1980s. In fact, planning and design for "JOP" started even before the Voyagers got to Jupiter. From its inception in 1977, I served on the imaging team for the Jupiter Orbiter. The team consisted of a dozen scientists affiliated with various separate universities or research centers. The project was soon renamed, appropriately, the Galileo mission. But then a series of disasters delayed and degraded the Galileo spacecraft, so that it only arrived in orbit around Jupiter late in 1995, where it would study the satellites for several years. By the end, in 2004, I had been on the imaging team for over a quarter century.

As Voyager 2 flew past Jupiter in July 1979, its trajectory swooped close to Europa. Remarkably, the key discovery that has explained nearly everything about the moons of Jupiter, although stimulated by the Voyager project, actually came before any spacecraft got to Jupiter. Only weeks before Voyager's arrival, theorists realized that tidal friction would be adequate to cook the rock of the moon Io and to melt most of Europa's water. Sure enough, the extraordinary technological and scientific achievements of Voyager included imaging of active volcanoes on Io and of the fresh, fractured crust of Europa.

NASA's usual strategy of planetary exploration is to follow such initial fly-by encounters with orbiters that can stay close to survey a planet in detail. The follow-up to Voyager was the Galileo orbiter. The bad news was that Galileo would still only be able to observe each moon during quick fly-bys, albeit now while orbiting Jupiter. The good news was that the brilliant engineering and operations of the mission allowed successful multiple encounters with each of the major moons, while overcoming major technological calamities that I will describe later in this book.

The images and other data received from those encounters form the basis for the interpretation of Europa's character that I develop in the following chapters. Looking at pictures is not enough. To *understand* what we have seen, we must consider the physical processes that could operate there. Deciding how those processes created what we see is crucial to assessing the likelihood of finding life there.

Before spacecraft arrived, Europa was assumed to be frozen within a thick, cold, solid, inert layer of ice. Now we know there is action on that surface and plenty of it. As I will describe in detail, thanks to tides, the icy surface is continually reprocessed at such a great rate that most of the observable terrain, structures and materials have probably been in place for less than 1% of the age of the solar system. By comparison, the surface of our own Moon formed early in the solar system's history and has changed little since then. The rate of change on Europa is more closely comparable to that on Earth. For example, during the time that Europa has been resurfaced, on Earth continents have been significantly rearranged, with North America moving away from Europe and India crashing into Asia. In the cosmically short time since dinosaurs became extinct on Earth 65 million years ago, the surface of Europa has been entirely recycled at least once, maybe twice.

During all that reprocessing, the surface ice has been bombarded by chemical substances potentially useful for life, from comets and asteroids, as well as from the swarm of Jupiter-orbiting particles. Just below the ice, the global, liquid-water ocean must contain materials released from the deep rocky interior, probably from on-going undersea volcanism.

The dynamic ice crust serves as the interface between the substances from the interior and those from exterior space. This barrier is solid enough to keep the chemistry in disequilibrium, but porous enough to allow interaction over various spatial and temporal scales. In broad terms, this physical and chemical setting seems to have what is needed to support life. A tidally supported ecology might be able to exploit it effectively, like the creatures I imagined at the beginning of this chapter.

On Earth, life tends to prosper at the boundaries of different physical regimes. Consider the diversity of life in a tide pool, at the land/sea interface. The natural flow of tidal water mixes the chemistry, allowing some organisms to commute between zones that satisfy their diverse needs, while others sit at anchor exploiting the flow, just as in our imagined view of the Europan crustal habitat.

On Europa, too, the geological activity that may provide the setting for a biosphere is driven by tides. These tides are enormous in comparison with terrestrial tides. On Earth tides are driven by the pull of the Moon, and to a comparable degree by the larger but far more distant Sun. Europan tides are driven by Jupiter, which is about as close to Europa as the Moon is to Earth, but Jupiter is 20,000 times more massive. Such tides have major effects:

- Tides distort the global shape of Europa on a daily basis, generating periodic *global stresses* that crack and displace plates of icy crust, driving a rich history of on-going tectonics and surface change.
- Tidal friction creates the dominant *internal heat* source, warming Europa enough to keep most of its thick water layer melted, maintaining the global ocean, and allowing frequent local or regional melt of the ice crust.
- Tides govern *Europa's rotation,* in ways that may enhance the tidal stress, leaving its imprint on the tectonic record of surface cracks.
- Tides control the long-term *orbital evolution* of several of Jupiter's largest satellites. A resonance among these moons, including Europa's, keeps the orbits from becoming circular, so they remain eccentric ellipses. In turn, the magnitude of the tides is directly dependent on the eccentricity. As the tidally-driven orbital changes modify Europa's eccentricity, all of the effects of Europan tides gradually change over tens of millions of years.

Orbital resonance, a global ocean, dramatic tectonics in the icy crust, and a daily ebb and flow of liquid from the ocean to the surface: These phenomena are all interdependent through the mechanism of tides. This interplay appears to have created a physical setting with all the ingredients and conditions for local habitable niches and for a long-lived global biosphere on Europa.

***O *

In this rich story, tides connect the orbits of Jupiter's moons to the geology of Europa, creating environments potentially suitable for life. But the unfolding of that story is, inevitably, interwoven with a story about life on Earth, about scientists and how they do science.

Making sense of Europa, both what we know and the limits of our knowledge, means understanding scientific logic and lines of evidence. But it also means understanding that science is a creative human enterprise. This scientific process is inevitably imperfect and uncertain, especially in a big-ticket scientific project, like a space mission.

It starts with the complex politics of deciding what to observe and how to do it, a process that determines how resources, prestige and influence are allocated in the scientific community. The jobs of designing, building, and operating the spacecraft (in other words all the things that go into acquiring the data) are largely handled by engineers, not scientists. Objectives are well-defined, so the process tends to be relatively objective and logical.

Then, with data in hand, scientific analysis begins. A descriptive phase comes first. Tens of thousands of images, in addition to all the other data, may come back from a spacecraft. Their essence is quickly reduced to written descriptions and power-point presentations. Illustrative examples are supposed to represent general observations, but they are at best anecdotal. Language is used to describe and digest what we observe, ideally in terms that retain its essential character without prejudicing subsequent interpretations. This descriptive phase is supposedly objective. However, too often, these descriptions reflect prejudices and personal interests of the project spokesmen.

Then comes the interpretive phase. Theorists construct "models" to explain what was seen. This modeling means considering an imaginary, artificial physical system that is simple enough to analyze, and then assuming that the results will apply to the real, complex planet under consideration. Occasionally the theorists have first-hand acquaintance with the data, but too often their models are simply inspired by the words used to describe the data to them. If the predictions of the artificial model compare well with the subjective description of reality, it is accepted that those physical processes were actually responsible for the character of the planet.

These scientific procedures depend on human judgment. What appears completely obvious to one scientist may seem questionable (or worse) to another. In the popular mind, science may be viewed as a systematic and perfectly logic-controlled process. In fact it is the usual mess of any endeavor built on human creativity and subjective judgment. Once the raw data are in hand and we begin to try to make sense of it, the scientific method is nothing like what you learned in school. Nevertheless, good scientists, and that includes the vast majority of my colleagues, try to be as objective as possible and to follow the evidence where it takes them.

Indeed several such lines of evidence, leading from the basic appearance of the surface through detailed modeling of the controlling processes, point to physical conditions that may plausibly support life. A tidal ecosystem exploiting habitable niches is consistent with what we have learned about the crust of Europa and the diverse ways that it changes over the daily cycle, over millennia, and over hundreds of millions of years. This book is about the evidence that has led to this picture of a permeable ice crust overlying a liquid water ocean.

This permeable-crust interpretation developed here runs counter to an alternative, heavily promoted view, in which an ocean, if any, lies isolated deep below a thick layer of ice. Of course, before the appreciation of tides in 1979, it was generally assumed that Europa's water was entirely frozen. To overcome skepticism about an ocean, those scientists considering tidal effects needed to show that even modest heating could keep most of the water melted. So low-ball heating estimates dominated many people's thinking, leading them to believe that the surface ice would be so thick that the ocean would be isolated below. In fact, it is completely plausible that there is enough heat to keep the ice quite thin.

Another factor that supported the isolated-ocean model was that the initial portrayal of the surface was assigned to politically powerful geologists on the Galileo team. They tended to describe Europa in terms of the most similar appearing analogs from their experience studying solid planets. Not surprisingly, these descriptions implied that the ocean played no direct role in producing the current surface. So powerful players (along with the theoretical modelers who had bought into their canonical descriptions), became invested in the notion of thick ice over an isolated, irrelevant ocean.

However, as I describe in this book, the preponderance of the evidence indicates just the opposite. Nearly everything we see on Europa's surface is the result of direct linkages with the ocean below. Most of the story is an intellectual one, logically combining theoretical modeling with observational evidence to infer what we can about Europa. But along the way the story is inevitably about the human enterprise, how we learn about a planet from data returned from a spacecraft.

Here is what I learned about science during my quarter-century with the Galileo mission: Big Science done in the context of a large space mission is governed by politics and money, as much or more than by the search for truth. In understanding the story of what we learned about Europa, that aspect of the human enterprise cannot be uncoupled from the intellectual one, because political power has been used aggressively to promote the isolated-ocean model.

It seems bizarre that political clout would be used to promote a scientifically weak position, but for those of us long involved in the very human enterprise of science, the situation is both familiar and disturbing. This book explains the scientific lines of evidence regarding Europa, but it also must address a party line inconsistent with that evidence. Deviation from that line is risky business. Selection for research funding, for publication in journals, and for academic jobs all depend on recommendations by other scientists, and that process is vulnerable to abuse. For me as a senior, tenured professor, the threat is not so serious. But those less secure may feel pressure to toe the party line or move on to other fields. Here is why it matters: The most brilliant young minds may leave science if they perceive it to reward something other than good research.

The political aspects of the story of Europan science play out largely within the scientific community. In this way they are different from the even more ominous attempts at political control of science from the outside, such as the recent effort to discredit the scientific consensus on climate change. There the corporate economic motives are clear. But what it is that drives politics regarding a moon of Jupiter is harder to grasp. Money and power must play central roles. Some of the players in the Europa story have very big offices, boss around a lot of helpers, and control huge research budgets, but it is still peanuts by corporate standards. So I have struggled to understand the motivations of such people. In any case, compared with the corporate world, the scientific community is a comfortably small pond where big fish don't need to worry about real sharks.

Perhaps the real currency here is ego and prestige. Long after I finished graduate school, my erstwhile PhD advisor, Irwin Shapiro, told me that what makes the scientific profession so strange is that we are competing for the admiration of our colleagues, with whom we are competing. That is a set-up for a perverted (or at best complex) social and political enterprise, certainly not a comfortable setting for someone with a thin skin. As it turned out, Europa's thin skin is what really made things uncomfortable for me.

***○ *

This book is mostly about Europa, but it is also about people and one of the remarkable things they do with their highly evolved abilities. Exploring the solar system, even a small part of it like the neighborhood of Jupiter, involves the coordinated efforts of a great many highly trained scientists and engineers with diverse talents and expertise. It should come as no surprise that the story of their joint effort features both selfishness and selflessness, egos in all shapes and sizes, gratification and frustration, and sophisticated systems of coordination and cooperation as well as competitive and self-promoting schemes that at times, to put it mildly, slowed down and even derailed the scientific effort.

As a participant in the decades-long Galileo project, I had a unique vantage point, formally an insider as a member of the imaging team, but always with the detached perspective of an outsider. For a variety of reasons, touched upon throughout the book, I never felt welcomed by the team. One reason was my scientific background. Most of the team was experienced in studying planetary surfaces or atmospheres, things you can see in pictures. Most of them had been involved with space missions together before. I was an expert in celestial mechanics, the mathematical study of the motions of planets and satellites. For most of the quarter century, I felt the rest of the team was wondering why I was there. Then, when Galileo finally began to return images of Europa to Earth, it became clear that my field was the key to understanding what we saw at Europa, and evident how significant those discoveries were. Scientifically, I found myself in the center of things, but the various social, political, and financial motives for keeping me marginalized were stronger than ever.

Several factors allowed me to stay in the game. First, Europa turned out to be arguably the most exciting subject of the Galileo mission's discoveries, and I brought unique expertise to interpreting what we saw. The premise of my original 1977 proposal—the one that got me onto the Galileo imaging team – proved to be more accurate than anyone (myself included) could have imagined: Most of what has shaped Europa involved tides, which could be best understood from the perspective of celestial mechanics.

Second, as Galileo approached Jupiter in the early 1990s, several remarkable scientists joined my research group. I like to think they were attracted by my dynamic leadership and towering intellectual reputation, but in fact it was surely the opportunity to participate in the Galileo project and to have access to the anticipated images that was their main professional motivation. Eventually about a dozen people contributed in important ways to our group's discoveries about Europa, as described in the following chapters, but the core group was my post-doctoral associate Paul Geissler and graduate students Greg Hoppa and Randy Tufts. As a returning older student, Randy brought decades of experience in the real-world political arena that was essential to keeping us afloat.

Third, my research group was fortunate to work at one of the premier centers of planetary science research, the University of Arizona's Lunar and Planetary Laboratory, a firm base from which to push back. And we had great facilities. LPL's Planetary Image Research Laboratory gave us immediate access to Galileo data as they arrived on Earth and to produce a range of essential image products. And, LPL concentrated a pool of remarkable people like those that accreted onto my research group.

So I had assembled the right group of people in the right place at the right time. We were ready with expertise and resources when the images came down. Ironically, without the mission's problems we never would have been able to play a central role. An antenna failure on the spacecraft (described in **Chapter 9**) reduced the number of images to a set that even a small group could handle, so Greg, Randy, Paul and I came to know every single image. Moreover, our relative isolation from the larger imaging team left us free to follow the evidence as we saw it, without pressure to conform to the constraints that governed so many others. Future investigations by ourselves and others will test our work, but nothing will top the thrill and satisfaction of seeing the pieces of the puzzle come together as we came to understand and to love Europa.

And, so, the story here has a satisfying ending. What we learned about Europa showed it as being more fantastic and exciting than we could have imagined, and our little group had the thrill of making great discoveries. But, like any "real-life" narrative for adults, the ending is not completely happy, nor even an ending at all. The study of Europa continues to this day, and is likely to continue for many decades. Europa is generally considered NASA's top objective for exploration in the outer solar system, and major spacecraft missions are under deliberation. Thus the stakes are higher than ever, and the inevitable maneuvering for funds, power, and recognition will continue to pervade the scientific process.

This book inevitably has villains, and although I often took their actions personally, I now realize that they were only following their nature. They are very effective at what they do. The people who did the some of the most misleading science described in this book have been selected to lead planning for future missions to Europa. Denigrating my work has been essential to keeping control of the political process and resources.

Fortunately, the story in this book is mostly about heroes, a huge number of people who worked together to make a robot in space perform miracles, even as bad luck kept throwing challenges in the way. I do make fun of the conservatism of the engineers and administrators of the project, but that criticism only applies where it was extended inappropriately to scientific analysis, discouraging creativity and risk-taking just where they were needed. In fact, without the engineers' and administrators' careful dedication, there would be no science to report. Those remarkable people got the robot to Jupiter and they made it work. The discoveries described in this book could not have been made without them.

I hope this book will convey some of the fun and excitement we had in unraveling the mysteries of Europa, and at the same time share a feeling for the rough and tumble of the modern scientific enterprise.

Touring the Surface

When Galileo discovered Europa in 1610, along with three other moons of Jupiter, his telescope showed them as only star-like points of light (**Fig. 2.1**). But even that low resolution was enough to show that the moons orbited around Jupiter, and to prove that the Earth is not the center of everything after all. Europa is about the same size as our Moon, but it is typically a couple of thousand times farther away from us. So over the following centuries, viewed from telescopes trapped on Earth, Europa remained only a single, unresolved point of light (**Fig. 2.2**).

Even so, important information came from that point of light. Its wavelength spectrum, what our eyes crudely see as color, depends on the material on Europa's surface. During the mid-twentieth century, instruments

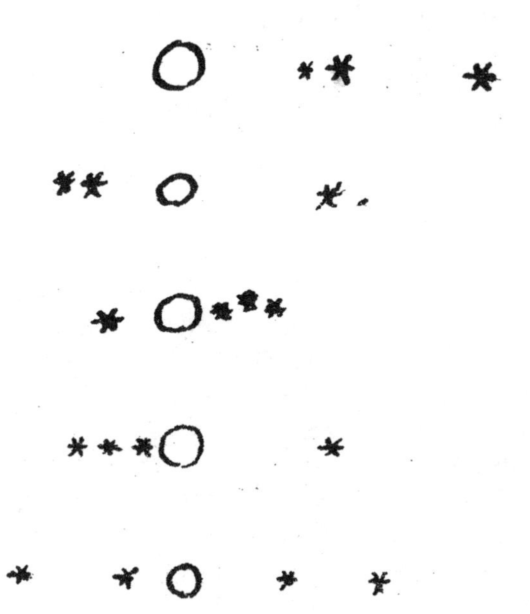

FIGURE 2.1: Galileo sketched the "Stars of the Medici," now known as the Galilean satellites. Each row shows the positions of the moons relative to Jupiter (circle) at a different time.

Figure 2.2: A recent Earth-based telescopic view still shows the satellites as star-like points of light, easily seen with a standard pair of binoculars. (Image by Michael Stegina and Adam Block of NOAO/AURA/NSF).

were developed that could accurately measure such spectra. Using a spectrograph and including measurements at infra-red wavelengths (too long for the human eye to see), in 1972 MIT graduate student Carl Pilcher inferred correctly that Europa's surface is predominantly water ice.

With the Voyager and Galileo spacecraft, we were able to carry cameras close enough to resolve Europa's surface in detail. These cameras had a resolution of 800x800 pixels, not much by the standards of modern consumer digital cameras, but compared with the single pixel views of Europa from Earth-bound telescopes, they were revolutionary. And we could get some great close-ups. These cameras had telephoto lenses, and occasionally the Galileo spacecraft swooped just over the surface. So the complete photo record includes broad global scale coverage, more detailed coverage of about 10% of the surface, and a few telling close-ups.

The images are spectacular, but even more exciting are the revelations of what lies below Europa's sunlit surface. For starters, subtle variations in the spacecraft trajectories, caused by Europa's gravity, constrained the moon's internal densities. There is a metallic core of radius 700 km, a rocky mantle, and an outer layer as thick as ~150 km with the density of liquid or solid water. Right at the surface, Europa's water had to be frozen because, without a blanketing atmosphere, any heat radiates rapidly into space before it could melt the ice. But below the surface, much of the H_2O layer is probably liquid, according to estimates of internal tidal heating. With such a thick layer of liquid water, this Moon-sized body has a global ocean comparable in volume to the combined oceans of the much-larger Earth.

The presence of large amounts of water surrounding a rocky interior had been predicted in the early 1970s by Guy Consolmagno, another MIT student, and John Lewis, his master's thesis advisor, based on theoretical models of how the satellites formed. According to these models, Jupiter's moons formed from a huge swirling cloud of dust and gas that surrounded the young planet, just as the planets themselves grew in a similar cloud around the Sun. As this Jovian "nebula" cooled, solid particles condensed out, like snow forming in a cloud on Earth. The particles gradually accreted together into satellites. The sequence of condensation determined the composition of each of the satellites. Then internal heating separated the materials by density within each satellite, with dense metal sinking to form a core, and lighter stuff rising toward the surface. A thick layer of H_2O on Europa was expected long before Voyager or Galileo got there.

As soon as the possibility of an ocean was raised, speculation about life began. There is some controversy about who suggested it first, but Guy Consolmagno thinks the issue was first raised in his Master's thesis. He is now a Jesuit brother, working at the Vatican Observatory, so he is reasonably credible. Plus, knowing Guy, I would trust him anyway.

What is it like down in that huge ocean under the ice? We know nothing about Europa's ocean floor, but a surprising amount of research has been based on speculation about it. Undersea volcanism has been discussed in detail because, when the ocean was believed to be isolated below the ice crust, volcanoes seemed to be the only source of heat and chemicals that might offer hope for life on Europa. The volcanic models were not motivated by any observations of Europa, but rather by recent discoveries of life supported by undersea volcanic vents on Earth and by the dramatic volcanoes on Europa's sibling Io. For Europa, there is no observational evidence either for or against volcanism or about anything else under the ocean.

We do, though, have plenty of observations of the surface. What we see in those images provides a key to understanding whether the ocean has what it takes to support life. The possibility of life no longer depends on speculation about under-sea volcanism.

We have excellent images of the surface of the ice because cameras on the Voyager and Galileo spacecraft had sensitivities and wavelength ranges that vastly exceeded those of the human eye. Even global views with low resolution (worse than 10 km per pixel) reveal much more than the uniform white water ice that would be visible to a human observer, as shown in **Color Plate 1.** At this resolution, dark (orange-brown) markings of still-unidentified substances on the bright background come in two major categories: splotches ranging from tens to hundreds of kilometers across and a network of narrow lines. The lines and splotches provide a first indication of what will prove (based on higher resolution and considerable analysis) to be the two major resurfacing processes on Europa: tectonics (the lines, which basically indicate cracking) and formation of chaotic terrain (the splotches, which are basically the result of melting). As we shall see, each likely involves direct interaction of the ocean with the surface, by cracking or melting, respectively.

Fig. 2.3 shows the disk of Europa from viewpoints increasingly eastward from (a) to (c). **Fig. 2.4** identifies a few basic features of Europa geography. The overlapping coverage includes areas in common on adjacent images, so the area on the west side of (c) appears on the east side of (b) for example.

The perspectives for these global views are illustrated schematically in **Fig. 2.5**. As Europa orbits around Jupiter, it keeps one face toward Jupiter. To do that, like the Earth's Moon, it spins on its axis with a period that matches its orbital period. This behavior is called *synchronous rotation*. (Actually, while the Moon's rotation is *exactly* synchronous, Europa's is merely *nearly* synchronous, but close enough for now. Later in this book, I discuss evidence that over the longer term Europa's rotation is actually non-synchronous, and that the whole surface slips around relative to the poles, but at least during this period of human observations the Jupiter-facing hemisphere has been nearly constant.) Synchronous rotation helps us define a latitude and longitude system, which we need to discuss locations on Europa. Latitude is easy to define; just as on Earth we measure it relative to the spin poles and equator. But how do we define longitude? On Earth, we define the prime meridian (zero longitude) as running through the Royal Observatory at Greenwich, England. On Europa, also by convention, the prime meridian runs through the point closest to Jupiter, which for practical purposes is fixed on the surface thanks to the synchronous rotation. Longitude 90° runs through the center of the *leading hemisphere* (the side in the direction of Europa's orbital motion).

The directions of the three global views in **Fig. 2.3** are indicated in **Fig. 2.5** by the arrows (a,b,c). For example, view (b) is centered near longitude 270° so when this image was taken, Jupiter was toward the left, and Europa was moving in its orbit *away* from the camera. Hence this side is called the *trailing* hemisphere.

A major landmark in the trailing hemisphere is the X formed by two globe-encircling dark lines that cross at a right angle slightly to the upper right of the center of the disk. This neighborhood is very important for several reasons. The two major types of geologic process are well represented here: The intersecting lines that form the X are global-scale tectonic features, and the dark patch in the lower crotch of the X, is Conamara Chaos, the archetype of what has come to be known as chaotic terrain. Most important, we have unusually complete photo coverage there. The detailed images of this area have given us great insight into the character and dynamics of Europa's active surface. At the same time, over-reliance on this site has produced some widely promoted generalizations about Europa

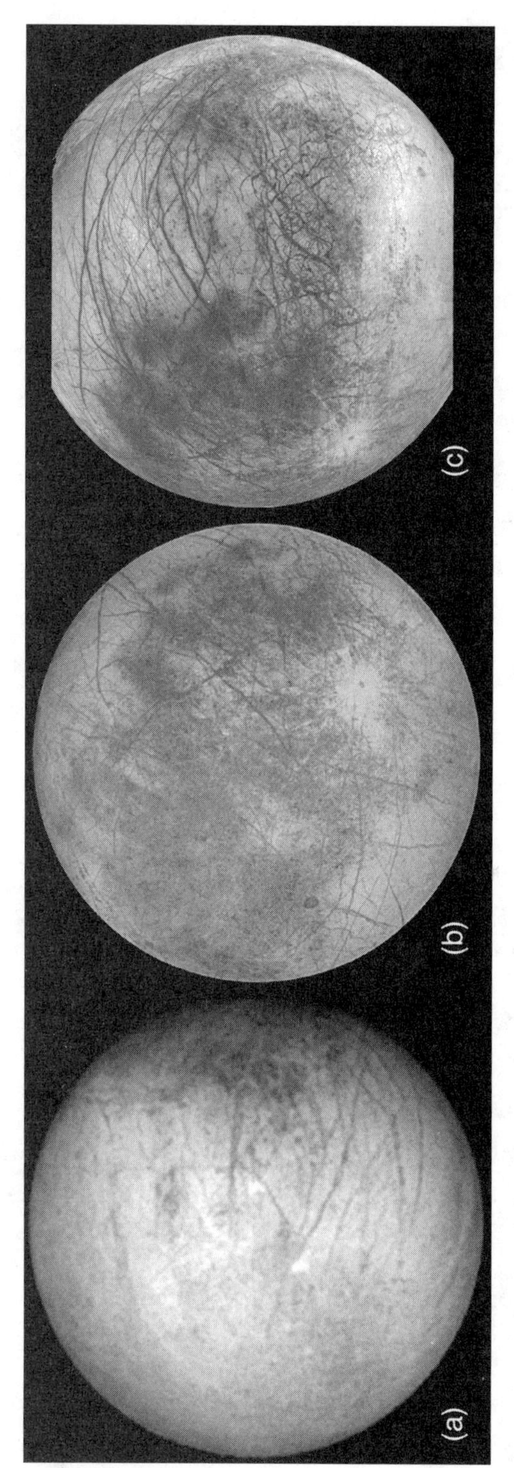

FIGURE 2.3: Three views of different faces of Europa with north at the top. Going from left to right, the three views are increasingly farther east. The overlapping coverage shows areas in common on adjacent images (see also **Fig. 2.4**). A color version of (b) is in **Color Plate 1**. The viewing perspectives are shown in **Fig. 2.5**. These global views have low resolution, with greater than 10 km per pixel.

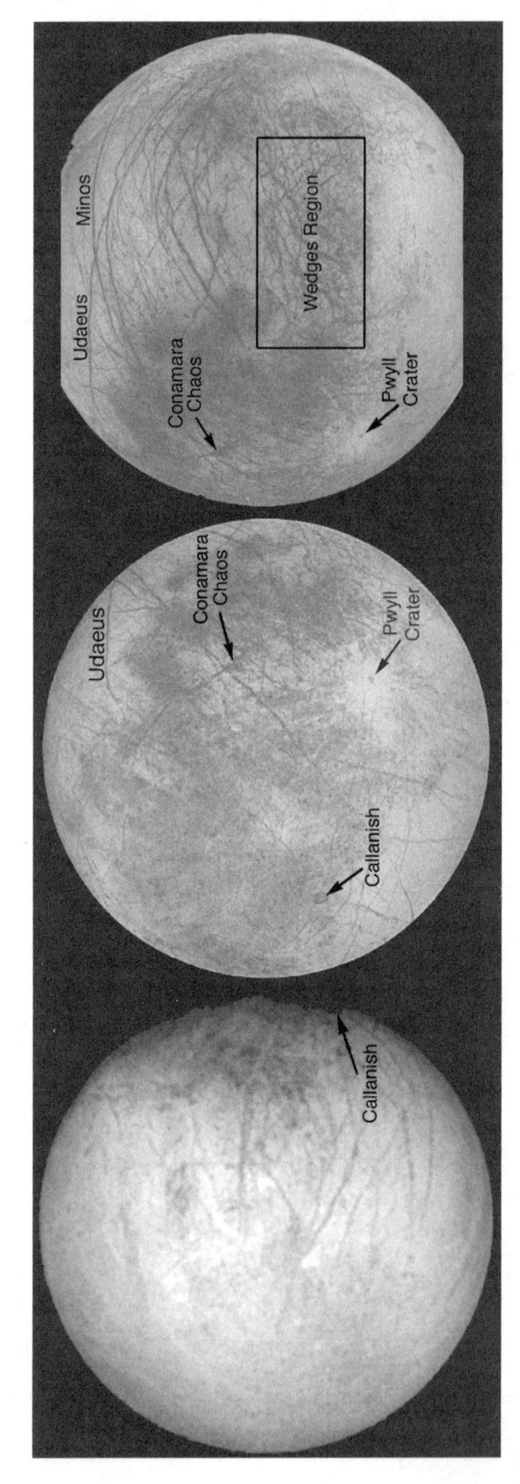

FIGURE 2.4: The views from **Fig 2.3**, showing some key geographical points of reference.

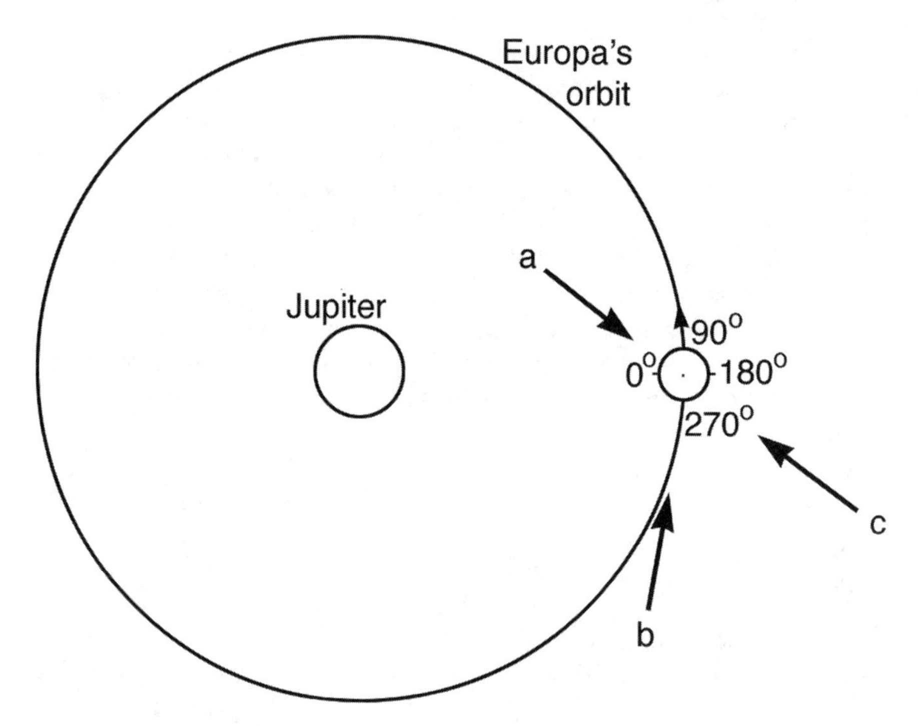

FIGURE 2.5: Europa, the small circle at the right, orbits around Jupiter. In this schematic (not to scale!), we are looking from the north, so the north poles are at the centers of Jupiter and Europa. The Jupiter-facing point is defined as longitude zero on Europa. The leading hemisphere (in the direction of the orbital motion) is centered at 90°, by definition. The directions of the views (a,b,c) in **Fig. 2.3** are centered at longitudes 40°, 290°, and 220°. Hence, **Fig. 2.3b** shows the trailing hemisphere, also shown in **Color Plate 1**.

that were not well justified, causing considerable confusion and misunderstandings. With appropriate attention to such pitfalls, the images of the Conamara area provide a key to what is seen across Europa.

The X at Conamara also serves as a convenient landmark for locating other images. The X lies just north of the equator and very close to the center of the trailing hemisphere at 270°. (**Fig. 2.3b** is actually centered on 290°.)

In addition to Conamara, some other geographical points of reference that are important later in the story are indicated in **Fig. 2.4**. The dark bands south of the equator in **Fig. 2.4c** are known as "the Wedges", because of their tapered shapes. The markings there divide the surface into distinctive rounded boxy shapes evident even on these low-resolution global views. Toward the north in **Fig. 2.4b** lie several long, dark, roughly east-west lines, which cross one another at oblique angles, two of which are

Udaeus and Minos Linea. The Udaeus-Minos region is important for several reasons: It contains these typical examples of global-scale (and even globe-encircling) lines; The tectonic features in this region played an important role in considerations of Europa's rotation; And (like the X at Conamara) these markings provide a point of reference for locating important sets of higher resolution images.

Wherever we look at the global scale, Europa is dominated by lines and splotches, that is to say by tectonics and chaos. Equally important as the ubiquitous splotches and lines is what is missing: craters. Unlike our own Moon, unlike the other icy moons of Jupiter, and indeed unlike most bodies in our solar system, Europa has hardly any craters. Apparently bombardment by small bodies has had a minimal role in shaping the surface that we see today.

Only a couple of impact features are evident at this scale. Crater Pwyll is about 1000 km due south of Conamara, appearing as the dark spot (about 20 km wide) in global views (a) and (b). That Pwyll is an impact feature is evident from the enormous system of bright rays of ejected ice that extend a thousand km or more in every direction. One ray even crosses the western side of Conamara Chaos. As a result of this whitening of its western side, Conamara, which is somewhat diamond shaped, appears to be more triangular in the global view.

Another prominent impact structure is the dark circle called Callanish near the left side of view (b) and at the eastern edge of (a), just south of the equator. This large feature (about 60 km across) does not seem very different from the splotches of chaotic terrain, viewed at this scale. What hints at its impact origin at this scale is its round shape. Higher resolution images confirm its impact origin, but they also suggest some very interesting similarities to chaotic terrain: Both may represent breakthrough to the liquid ocean.

Why are there so few craters? The answer is probably the same as for the Earth: The planet has undergone active geological resurfacing that has erased the record of nearly all impacts. The current surface must be less than about 50 million years old, or else it would show far more craters than it does. We know of no major event that wiped the entire surface clean all at once about 50 million years ago. Instead, the continual, gradual resurfacing manifested by chaotic terrain and tectonics provides a ready explanation for the youthful appearance of Europa.

To summarize then, even in these three low-resolution global views, the main points of our story are introduced. Europa's appearance is dominated by features that represent the effects of resurfacing that has been rapid and recent, and that is probably on-going. The global view shows the two major categories of the resurfacing processes: tectonics and chaos formation. As I show later, these processes not only modify the surface, but they generally provide, in diverse ways, access between the surface and the liquid ocean. The dark material that marks the sites of these processes is indistinguishable from one site to another – it looks the same where it marks linear tectonic features or chaotic terrain. This similarity probably reflects a common feature of these process: They all involve interaction of the surface of the ice with the ocean. The dark markings may represent concentration of impurities as ice was warmed by nearby liquid water, or they might simply consist of recently exposed substances from the ocean that lies just below the ice. The implication seems inescapable: The surface features of Europa suggest a dynamic, on-going connection between what lies above and what lies below.

***O *

Global images were a revelation, but the best was to come, as Galileo made closer passages to the surface and the images zoomed-in with increasing detail. **Color Plate 2** shows a 1000 km-wide portion of the Udaeus-Minos region, a relative close-up compared with the global views. Many of the scratchy dark lines resolve into double lines, typically about 10 km wide, with a bright lane between them. Planetary geologists have given these double lines the name "triple bands". They counted the central lane between the dark lines, but the central lanes are not brighter than the surrounding surface. Another quirk of this terminology is that, on Europa, the term "band" is more generally used to describe dilated cracks, where adjacent plates have pulled apart from one another. As we shall see, the only thing that triple bands and dilational bands have in common is that both initiated as tectonic cracks.

In this region, at this scale, we also see a great many dark spots, typically about 10 km wide, and quite common in this region of Europa. During the early part of the Galileo mission, the International Astronomical Union, the organization charged with naming things in space, deliberated at length about what, or even whether, to name this class of feature. My doctoral student Randy Tufts called them freckles, but astutely recognizing the cachet of Latin in academic circles he proposed the term *lenticulae*, which means the same thing. The spots got their official name.

As we shall see, most of these spots represent very small patches of chaotic terrain. One reason so many of them appear to be about 10 km wide is that such patches usually have a dark several-km-wide halo around them. That typical halo size is independent of the size of the patch of chaos. So no matter how small a patch of chaos may be, at this resolution it may appear to be about 10 km wide. This effect explains why so many spots are about 10 km wide, even though the chaotic terrain has no preferential size. There has been a lot of confusion caused by misuse of the term *lenticulae* (it is a big deal in **Chapter 14**). The key is that it was defined to describe these dark spots as they appear on low resolution images, where brightness variations could be seen, but not surface relief (or "morphology").

The similar, but larger, splotches also prove at higher resolution to be chaotic terrain. Amazing confusion has ensued from the practice of giving a single kind of terrain different names depending on the resolution at which it was imaged. When those larger splotches were seen with low resolution (starting with Voyager), they came to be called "mottled terrain". Like lenticulae, the term mottled terrain really was introduced to describe darkened patches on the surface. So a patch of "mottled terrain" is the same as a lenticula, only bigger. But then at higher resolution, where the surface relief can be seen, mottled terrain (like most lenticulae) correlates with chaotic terrain. In this book I use the term chaotic terrain because it is based on better and more complete imagery than was available when the term mottled terrain was invented to describe low resolution pictures.

Another zoom in from the global scale shows the broad area ranging from the big X at Conamara all the way south to the impact crater Pwyll, about 1000 km away (**Color Plate 3**). Compare this image with the global view in **Color Plate 1**, for example. Again, the global scale lineaments (the lines of the X, for example) are seen to divide into double dark lines at the higher resolution. (See also **Fig. 2.6.**) We can also see more clearly the ray pattern of material splashed out from the impact at Pwyll. Conamara Chaos, nestled into the crotch of the X is somewhat diamond shaped, 80 km across, with a wisp of bright ejecta crossing its western side.

***O *

As much of an improvement as the kilometer-resolution images were over the global views, there is still something frustrating about them. We cannot quite tell what it is we are seeing. Part of the problem is that, to enhance information about the reflectivity (and colors) of the surface material, those images were generally taken with direct sunlight illuminating the surface from nearly perpendicular to the surface. But that illumination hides the morphology (relief, or shape) of the surface. There are no shadows, no variations in brightness due to slopes, few clues regarding the structure of the surface. We cannot see any hills, ridges, pits, valleys, gorges, cliffs, or any of the other landforms that might appear on a planetary surface. We have seen in the images above markings corresponding to all of the important geological features and processes, but no direct evidence for them yet.

To be sure, part of the explanation is that there simply is not much topography on Europa. Hardly any features vary by more than a couple of hundred meters from the mean elevation, and slopes are very gradual. Europa is one of the smoothest bodies in the solar system. But to really understand the geology, we need to look at higher resolution images, and with sun angles low enough to reveal the *bas-relief* of the subtle surface structure.

Galileo imagery includes coverage of about 10 percent of the surface at resolution of 200 m/pixel. We would have had much more coverage, were it not for the failure of the spacecraft's high-gain communications antenna (described in **Chapter 9**). But these images were planned with a geological survey in mind, so the illumination is generally good for revealing the shape of the surface. In fact, the high quality of these images does exacerbate our frustration that they cover such a small fraction of the surface. On the other hand, the images we do have, combined with the broader context seen at lower resolution, suggest that we may have sampled most of the major types of terrains, structures, and processes.

The Conamara area is shown at this higher resolution with the more oblique lighting in **Fig. 2.7**. With this illumination, the topography is evident, where it was missing entirely in **Fig. 2.6** for example. (Of course, when viewing an unfamiliar type of object, the brain can produce optical illusions. Sometimes, hills can look like holes or ridges can appear to be troughs. It may help if you bear in mind that in this image the lighting comes from the right, as it does in most of the images that show surface relief in this book.)

In **Fig. 2.7** the true character of the lines and splotches is revealed: The large-scale lines prove to be complexes of ridges. Chaotic terrain proves to be the site of disruption, where blocks of older surface have been separated and displaced, like rafts of ice within a melted patch of lake ice on Earth. The bright ray of ejecta is still apparent running north-south across the western side of Conamara. Structural details of lenticulae are visible: Three dark spots north and northeast of Conamara in **Fig. 2.6** (and even visible in the global images) are resolved in **Fig. 2.7** into small patches of chaos each with a dark halo, detail that was not evident at lower resolution. Where there is no chaos, the terrain proves to be densely covered by mutually criss-crossing ridges ranging in width from about a km wide down to as narrow as the image can resolve.

Even this brief first look at Europa's surface touches on many of the issues of Europa science detailed in subsequent chapters. Most broadly, Europa is

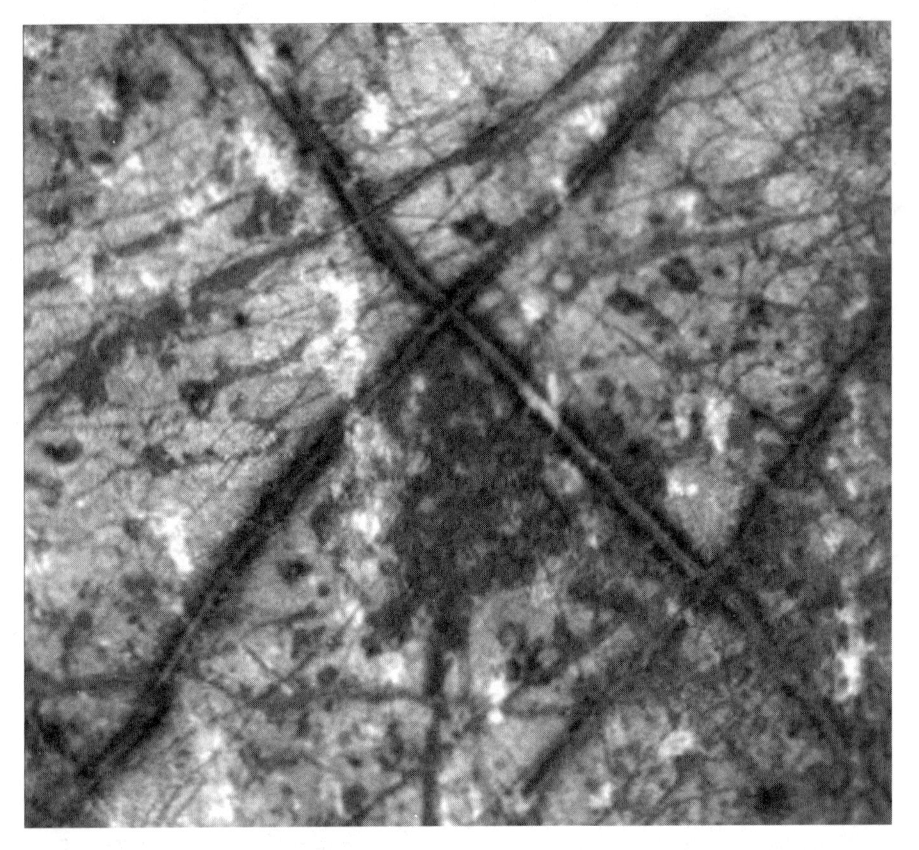

FIGURE 2.6: A black-and-white enlargement of a 300-km-wide portion of Color Plate 3, showing Conamara Chaos (80 km wide) and the area around it.

dominated by two classes of terrain, tectonic and chaotic, which correspond respectively to the dominant effects, stress and heat, that operate on the icy crust. Some of the most dynamic resurfacing involves formation of ridges along cracks and of chaotic terrain. So to complete this introduction to what we see on Europa, I focus on the appearance of ridges, and then on chaotic terrain.

The major ridge complexes that make up the global-scale lineaments (like the big X at Conamara) each consist of multiple pairs of parallel ridges. The pairs are roughly parallel to one another, although crossing or intertwined in various places. The ridge complexes are fairly bright, at least as bright as any of the surrounding terrain. Surprisingly, the darkening that characterized these global scale lineaments is hardly evident here at all.

Why did the so-called "triple bands" disappear in this picture? If we look carefully at **Fig. 2.7**, we see that along either side of the major ridge

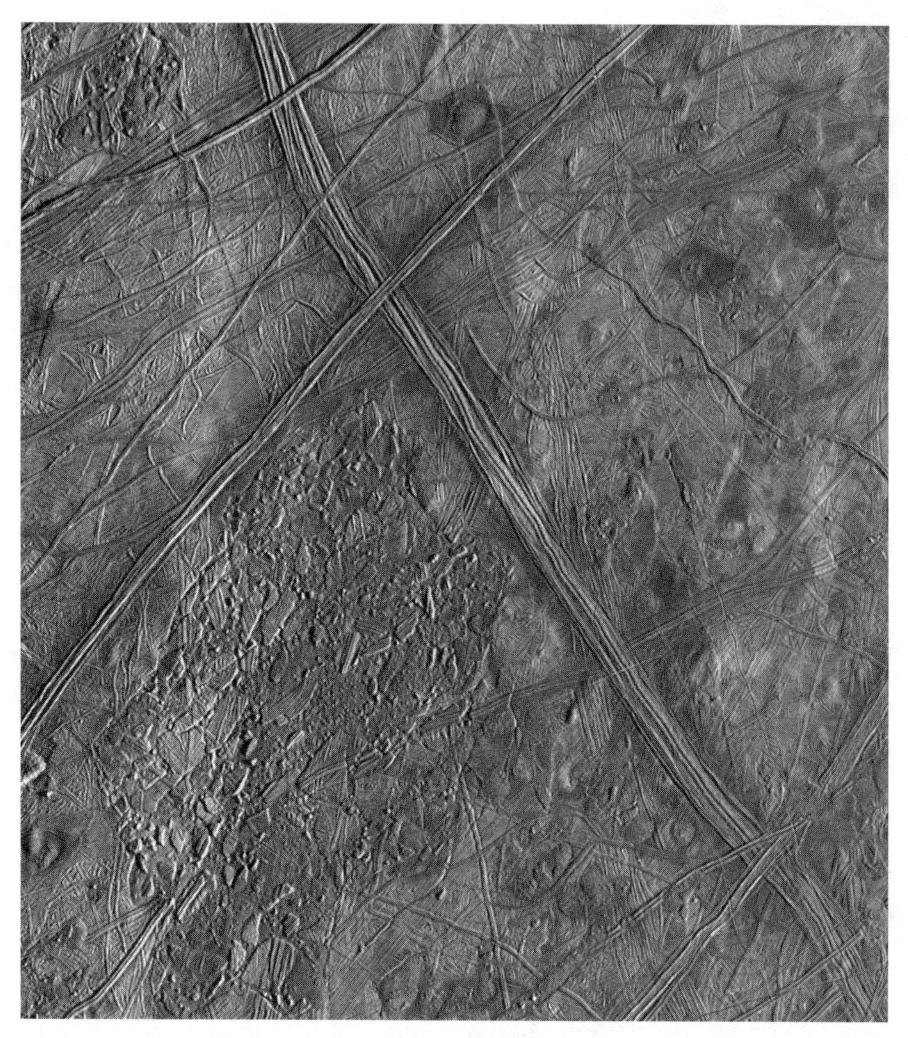

FIGURE 2.7: The Conamara Chaos region in a 180-km-wide mosaic of images at about 180 m/pixel, showing the character of the chaotic terrain. Conamara fills much of the lower left quadrant of this mosaic. Illumination is from the right. Global lineaments are complexes of double ridges. The darkening that dominates their appearance at low resolution is only a diffuse effect along the margins of these ridge systems. The nearby *lenticulae* prove to be small patches of chaotic terrain with their similar dark margins.

complexes the adjacent terrain is slightly darker than average. It is this subtle, diffuse darkening that shows as the dark lines on the contrast-enhanced global scale images or as the double dark components of the "triple bands" that we have seen at intermediate resolution (e.g. **Fig. 2.6**). The ridge systems, which now are revealed to be the most significant structure of the global lines, lie between the faint double dark lines. In the lower resolution images they simply appear as the relatively bright center line of the triple bands.

The dark coloration on the margins of the ridge complexes, which seemed so prominent on the lower resolution images, proves to be barely recognizable in higher resolution images with oblique illumination (**Fig. 2.7**). These darkened margins themselves do not directly mark any topographic structure, except that the surface seems to be somewhat smoother and lower where it is darkest. The significance of the darkening is in the correlation of the coloring agents with ridge complexes: The ridges probably mark cracks along which ocean water has been able to reach the surface; thus, these margins are the first of several examples of darkening (and perhaps smoothing) associated with oceanic exposure.

Major ridge complexes (like those in **Fig. 2.7**) are composed of sets of double ridges. In fact, nearly all ridges on Europa come in pairs of identical components, each pair remarkably uniform along its length.

The terrain surrounding Conamara consists largely of criss-crossed double ridges. With one of the rare very-high resolution images (here at 21 m/pixel), we can zoom in on an example (**Fig. 2.8**). With Galileo's limited picture budget, only isolated sites have been observed so well, but they reveal the character of the types of terrains. In **Fig. 2.8** we see rather typical densely-ridged terrain. The previous surface has been covered repeatedly by newer ridges, each crossing what was there before, until nothing is visible but ridges on top of ridges. The largest ridge-pair crossing this field of view is about 2 km across and about 100 m high, about as large as any ridge gets on Europa. Among the youngest features shown here are very fine cracks crossing and cutting older ridges. The double ridges may have formed along the borders of such cracks, while the recent cracks may not have had time for ridges to form.

This type of densely ridged terrain is sometimes called "bright plains" or even worse "background plains", both misnomers based on low-resolution images, in which the areas looked smooth because the ridges were not visible. Even worse, the term "background plains", as used by the geological mappers of the Galileo imaging team, promoted a misleading implication that it is the oldest type of terrain on Europa. That terminology also suggests that the youth of Europa's surface is due to some sort of slate-cleaning event, which produced a background starting condition on which all subsequent geological processing acted. However, there is no evidence that resurfacing has ever been other than a gradual process of continual renewal by a fairly constant set of processes.

As higher resolution images showed the true character of this terrain, the

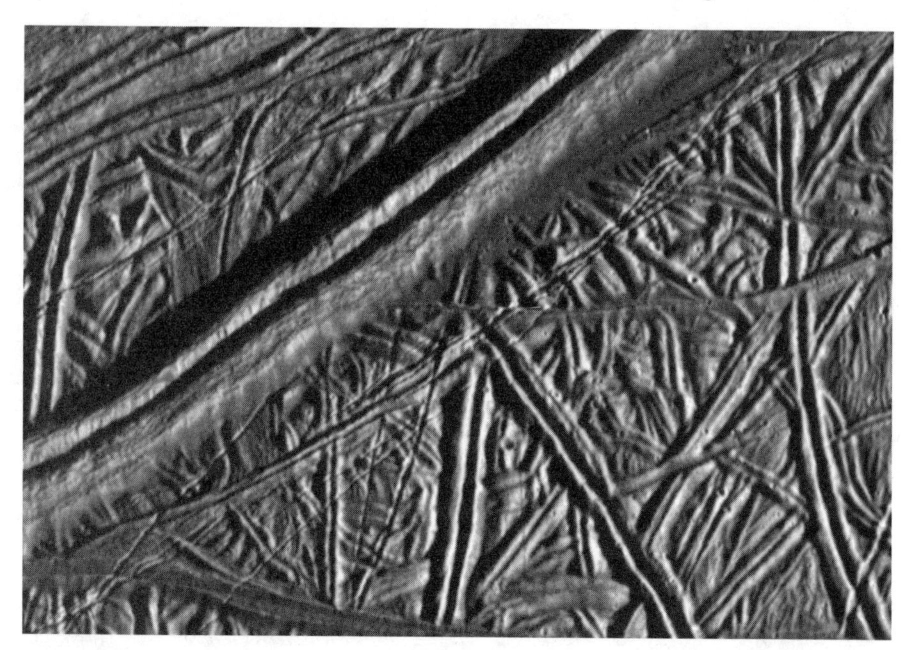

FIGURE 2.8: A 12-km-wide portion of a very-high-resolution image (21 m/pixel) taken during orbit E6. This image was intended to examine ''bright plains'', which proved to be densely ridged terrain. This area is near the northern edge of **Fig. 2.7**. Ridges on Europa usually come in pairs.

earlier terminology became obsolete. Terms like "bright plains" and "background plains" should have been abandoned, along with "mottled terrain" and "lenticulae". Unfortunately their use has continued, resulting in a confusing taxonomy. In general, classifying a single type of terrain in multiple ways depending how it looks in different pictures is a bad idea. In Galileo science analysis, this practice has caused confusion, misunderstandings, and incorrect generalizations and inferences.

The common denominator for all ridge systems on Europa is that ridges come in pairs. Contrary to a too-common misconception, the double ridges do not correlate with the double dark lines of "triple bands" that were seen at kilometer-scale resolution. That confusion may stem from an early Galileo press-release montage (used in numerous review talks by Galileo spokesmen) that showed double ridges blown up to match the size of a triple band displayed next to it. I want to emphasize the correct relationship between ridge pairs and triple bands: Ridges are fairly bright and each is at most about 1 km wide. The dark margins are something else entirely. They are typically ~10 km wide, they are found outside the area of the ridges, and they are only associated with multi-ridge complexes (like those that make-up the

X in Conamara) or with a few of the largest ridges. The double dark lines are only pale diffuse markings, while the double ridges, as a class, represent one of the most important and telling morphological features on Europa.

Ridges are important because they cover so much of the surface. But they also serve as a marker of one of the most fundamental geophysical processes at work on Europa: the cracking of the crust. Ridges seem to correlate with tensile cracking of the icy crust, according to the following train of logic: The global and regional lineaments correlate reasonably well with theoretical tidal stress patterns. Because these lineaments seem to comprise complexes of double ridges, it is reasonably assumed that simple double ridges are similarly associated with cracks. What is more, the double nature is due to ridges running along each side of a crack.

So tracking the ridges on Europa gives us a way to unravel the history of the cracking of Europa's crust. From the paths that the ridges take over the surface, we can read the history of the stress of tides. Some characteristic patterns even provided our first geological evidence that there is indeed an ocean down below the thin ice shell. Tracking these *tidal stress* patterns will be a big part of the story to follow, but first let's take a closer look at what *tidal heat* can do to the surface.

Conamara Chaos is fairly typical chaotic terrain. We have already seen it in a sequence of ever higher-resolution images. Going a step further, a set of high-resolution images span a belt across the southern half of Conamara, showing much greater detail, including the portion in **Fig. 2.9**. Throughout Conamara the surface appearance suggests thermal disruption, leaving a lumpy, bumpy matrix with somewhat displaced rafts, on whose surfaces fragments of the previous surface are clearly visible. The rafts seen in detail in **Fig. 2.9** also appear in **Fig. 2.7** in their broader context.

Chaotic terrain looks like a place where the crust briefly melted, allowing blocks of surface ice to float to slightly displaced locations before refreezing back into place. Similar features are common in seasonally frozen lakes or in Arctic sea ice, where the underlying liquid has been exposed.

Formation of chaotic terrain clearly represents the destruction of an earlier surface. In the case of Conamara, that earlier surface was tectonic terrain. The surfaces of the rafts still display fragments of a terrain that was covered with ridges and cracks, essentially the same terrain that immediately surrounds Conamara. Like pieces of a picture puzzle, the rafts can be readily reassembled into fairly continuous areas, generally reconstructing a few of the major ridge systems that crossed the region. However, the entire destroyed surface cannot be reconstructed, because most of it has been

FIGURE 2.9: A very high resolution (54 m/pixel) image of detail within Conamara Chaos.

broken into lumps too small to show their earlier surface or too melted to retain their original shape.

While formation of chaotic terrain has destroyed the previous tectonic terrain by breaking it up and melting much of it, we can also see in **Fig. 2.9** the revenge of tectonics. After Conamara formed and refroze, more cracking has occurred. We can recognize cracks that formed after Conamara Chaos, because they cut through the lumpy matrix and when they reach rafts they either slice across them or wend their way among them. They contrast with those older tectonic features that lie only on the rafts and do not extend into the matrix. One example of a post-Conamara crack runs diagonally across the lower left corner of **Fig. 2.9**. Another toward the upper right snakes its way among rafts, along the northern edge of one of the larger rafts, and across the lumpy matrix. This example has already begun to form a double ridge.

In this way cracking and ridge-building have already begun to resurface Conamara. In other places on Europa, chaotic terrain has been covered by ridges to varying degrees, including many cases where the chaotic terrain is barely recognizable under the criss-crossing ridges. Ridge formation seems to be as effective a resurfacing process as chaos formation. In **Fig. 2.8** we saw terrain where ridges had covered other ridges, and in **Fig. 2.9** we see where ridges have begun to cover chaos. Ridges cover over what was there before; chaos formation destroys it.

Conamara is an unusually fresh example of chaotic terrain. For that reason it stood out so prominently in earlier pictures that it was selected for targeted high-resolution study. As a result, it became the archetype for chaotic terrain. After that, the characteristics of this specific example of chaos were assumed to apply in general, so chaotic terrain was described as rare and recent. This misconception, a classic observational-selection error, became one of the widely canonical results of Galileo imaging.

In fact, most chaotic terrain is much older than Conamara and modified by subsequent resurfacing. The older examples of chaos are simply harder to see, which led to the false impression generated early in the Galileo mission that chaos is a relatively recent phenomenon. On the contrary, like tectonic processes, formation of chaotic terrain has occurred throughout the geological history of Europa, as far back as the record goes.

Conamara is only typical as an example of very fresh chaos, not of Europan chaos in general. It must have formed recently relative to the rest of the surface. Given that the surface of Europa is less than 50 million years old, according to the lack of craters, Conamara itself probably formed within the past million years or so.

Fig. 2.9 also shows several tiny craters. Two examples (each a couple of hundred meters across) lie on a raft near the far left. Based on relationships with other craters, these are probably part of the large population of tiny secondary craters, formed by debris ejected by the rare larger impacts on Europa.

Returning to chaotic terrain, not only is Conamara not typical of chaos age, at 80 km wide it is not typical of the size of chaos patches either. The largest single patch that we have examined in reasonable resolution images is roughly circular and about 1300 km across. Other even larger chaos regions are evident as the dark splotches in the global images (**Fig. 2.3**). The distribution of sizes of patches of chaos is such that the smaller they are, the more there are, down to sizes so small they are barely recognizable. Evidently the numbers are limited only by the fact that we do not have adequate imaging to see the smallest patches.

A small patch of chaos is shown at the top of **Fig. 2.10**, a blow-up of the area just above the X in **Fig. 2.7**. Here the lumpiness of the texture is finer in proportion to the small size of the chaos area. There are no rafts large enough to reveal older terrain.

The patch of chaotic terrain in **Fig. 2.10** appears to be bulged upward. In some interpretations, features like this one are taken to be upwelling of magmas (slush or viscous ice) that rose and spread over the surface. However, such bulging would also follow naturally from exactly the same sort of fluid exposure as appears to have created Conamara and other

FIGURE 2.10: A blow up of the top-center portion of Fig. 2.7, showing a bit of everything: tectonic terrain, densely filled with double ridges (which appear as "background plains" at low-resolution); ridge complexes with dark margins (which appear as "triple bands" at low resolution); and a small patch of chaotic terrain with similar dark margins (which appear as a typical *lenticula* at low resolution). The area shown here is about 50 km across.

chaotic terrain: After the melt-through from below, buoyancy would bulge up the surface during subsequent refreezing (**Chapter 13**).

The dark diffuse halo around the small patch of chaos in **Fig. 2.10** is significant, because it explains the appearance of such features at low resolution, as well as the source of a crucial misconception: that the size of

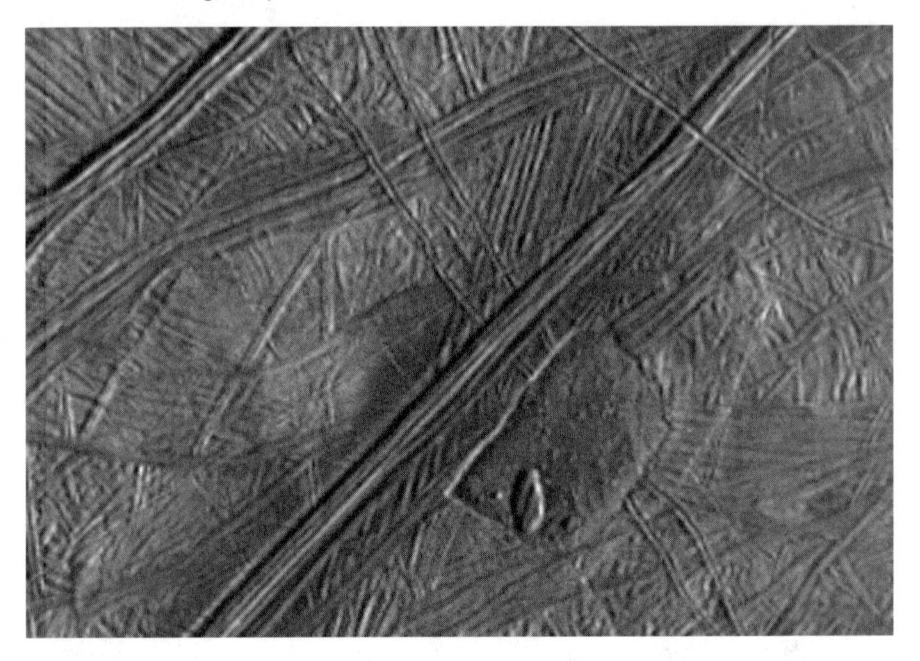

FIGURE 2.11: A small patch of typical chaotic terrain (~10 km across) lies surrounded by densely ridged tectonic terrain. Despite its small size, it contains a raft with a bit of the older ridged terrain. Patches of chaotic terrain are found at all sizes down to the limits of resolution in our images.

lenticula (the low-resolution spots) can be used to infer the sizes of patches of chaotic terrain. This halo shows up in **Fig. 2.6** as a lenticula, just as the similar darkening along the major ridge complex (that runs diagonally across **Fig. 2.10**) shows as double dark lines (forming a so-called "triple band") in **Fig. 2.6.** The scale, the amount of darkening, and the diffuse appearance are nearly identical whether the darkening is a halo around a very small patch of chaos or a pair of dark margins along a global ridge complex.

Once the relationships between appearance in global- to regional-scale images and appearance at this higher resolution are understood, we can interpret the true meaning of the archaic taxonomy developed for the earlier, low resolution data: the splotches (or "mottled terrain") are chaotic terrain, "triple bands" are major ridge complexes with diffuse dark borders, "bright plains" are densely ridged terrain, and "lenticulae" are small patches of chaotic terrain, enhanced in size by their diffuse dark halos.

Europa's surface is young, and it appears to be continually resurfaced by two dominant processes: tidal stress forming cracks, ridges, and related tectonic features, and tidal heat creating chaotic terrain. The area sampled in **Fig. 2.11** encapsulates that fundamental character: densely ridged terrain

Sumo on Ice

FIGURE 2.12: Chaotic terrain is most likely due to melt-through from below, but the explanation suggested here is interesting. In any case, the appearance of chaotic terrain strongly suggests areas of exposure of the liquid water that is ordinarily below the ice.

surrounding a small patch of chaos with all the appearance of a melt-through site, including a raft of displaced ridged terrain. While this patch of chaotic terrain is similar to that in **Fig. 2.12**, we can probably rule out Sumo as a major process on Europa.

Wherever it occurs, each of the dominant processes, stress and heat, wipes out what was on the surface before. This resurfacing is rapid and recent. Each appears to involve the interaction of the liquid ocean with the surface. These processes and the conditions they produce may create and maintain a variety of habitable niches capable of supporting life. As we shall see, all of this activity is driven by tides. And as we shall see too, despite strong evidence for this picture of Europa, the scientific process itself has proved to be more Sumo-like than I ever expected.

Doing Science

If what we were discussing were a point of law or of the humanities, in which neither true nor false exists, one might trust in subtlety of mind and readiness of tongue and the comparative expertness of writers, expecting him who excelled in these qualities to make his arguments the most plausible; and one might judge that to be correct. But in the physical sciences, where conclusions are true and necessary and have nothing to do with human preferences, one must take care not to place oneself in the defense of error.

—Galileo, 1632

Our tour of Europa in the previous chapter suggested that the important surface features, and the processes that formed them, all involved an ocean linked to the surface through a thin, permeable crust of ice in a variety of ways: Narrow cracks allow ocean water to flow to the surface and form double ridges. Cracks have often spread open, filling with fresh ocean water. From time to time, and place to place, local heating has thinned the ice, often to the point of exposing the ocean and creating chaotic terrain.

The images themselves do not show those things happening. I would love to have a video camera monitoring Europa, and something like that may happen in the future, but for now we only have snapshots. Those images don't show heat, they don't show ocean water reaching the surface, they don't show plates moving. Although those images were the culmination of an amazing engineering project, they really were only the start of the science. Most people seem to think of science as a body of knowledge, not surprising because that is largely how it is presented in school. But really, science is a process, the thing that scientists do. Doing science is what gets us from the images sent by a robot camera to understanding the character of Europa.

So when I describe the melting, the cracking, and the multiple ways that ocean water reaches the surface, I am reporting what we learned by doing science and following various lines of evidence and inference. This process led me and my students and colleagues to this permeable nature of Europa's icy crust.

But, if science is a process, it is a human one. There is rarely complete agreement, and often bitter controversy. In the case of Europa, a very different interpretation from ours has been widely promoted. In that view the ocean is completely isolated from the surface, and has been essentially irrelevant to the character of what we see in the images. Instead, all the activity occurred within a thick layer of ice. The idea is based on the fact that, given enough time, even solid materials can flow. On Earth the movement of glaciers shows that solid ice flows easily. Even rock has deformed into seemingly improbable shapes over the aeons. So according to the isolated-ocean model, the chaos and ridges and everything else that we see on the surface of Europa resulted from low-density blobs of solid ice flowing up through the crust. The problem is that the logical chain from observations to the isolated-ocean model has, on close inspection, weak or missing links.

Nevertheless, the isolated-ocean model was widely reported and accepted during much of the Galileo mission's time in orbit around Jupiter. Thanks to the peculiarities of the power structure within the scientific enterprise, there remains a strong impetus for those beholden to the NASA space-mission establishment to pay homage to that model. On the other hand, recognition of the direct role of the liquid ocean in shaping the surface has developed more slowly and recently. It seems to be widely accepted, even obvious, to scientists who are able to operate more independently or are free to express their views, and even the hard-line isolated-ocean people are starting to hedge their bets. But for me and my students, it has been a long hard fight.

Where did the isolated-ocean model come from? To a large extent, it was the result of history. Until 1979, no one suspected that water on Europa would be anything but solid ice. Then, only a few days before Voyager's arrival at Jupiter, theorists predicted substantial heating of the Galilean satellites by the friction of tides. The timing was perfect because the Voyager images immediately showed active volcanism on the satellite Io, which could only be explained by the tidal heating. I describe in the following chapters how tidal heating works, but the bottom line is that on Europa the tidal action is adequate to keep nearly all the water in a liquid state, except for a thin layer of ice at the surface. However, at the time of Voyager, any liquid water so far from the Sun seemed a radical notion. To make a credible case, the theorists needed to show that a liquid layer would be possible with even a minimal amount of tidal heating. Thus the scientific literature was dominated by conservative estimates of heating rates, allowing only a thin liquid layer under a thick ice crust.

The thickness of the ice layer on top of the ocean depends on the rate of internal heating. We know the surface is kept cold (about −170°C) as heat radiates into space, but tides keep generating heat inside Europa. Most rigid materials transport heat by conduction. It is conduction that carries heat through a pan from your stove to your eggs. If the same is true for the ice on Europa, then the faster that heat is produced in the interior, the thinner the ice must be to carry away the excess.

In fact, the thickness of the ice adjusts itself according to the heating rate. If the ice were too thick, extra heat would build up inside, melting the bottom of the ice until a balance was reached. If it were too thin, excess heat would radiate away quickly and the ice would thicken. For the highest imaginable heating rates, each 10-meter-square patch of ice would conduct out heat equivalent to a 50-watt light bulb. In that case the ice would adjust itself to one or two kilometers (roughly a mile) thick. More modest heating rates would mean proportionately thicker ice. So, for example, 10 watts generated under the same area would yield ice thickness of 5 to 10 km (roughly 5 miles).

So we can see that conservative estimates of the amount of internal tidal heating would tend to imply relatively thicker ice. That trend is even stronger if you consider the possibility that heat could be transported through the ice by convection instead of conduction. Convection usually operates in fluids, such as our atmosphere in turbulent weather, when warm air rises and carries up heat, while cool air sinks down. Solid ice can flow too, so convection could be possible within Europa's crust, speeding the transport of heat. Such convection would not kick in unless the ice were thicker than about 20 km. If it did, then the ice could be thicker than it would be with simple conduction. So if you are conservative in your guess about the tidal heating rate, you can end up computing that the ice is tens of kilometers thick.

In fact, such a thick crust could transport the heat almost as fast as a thinner layer if the ice were convecting, instead of just conducting the heat. Remember, even solid ice can flow, albeit slowly. For convection within the solid ice crust, warm ice would flow up and cooler ice down, transporting heat like a conveyor belt. If the ice crust were 20 km thick or more and *convecting*, it could transport heat as quickly as a crust less than 10 km thick *conducting* the same amount of heat. For a given heating rate, estimates of the ice thickness depend on whether it is convecting or not.

Theoretical considerations suggest that convection may require the ice to have very special material properties, such as just the right grain size. We can only speculate whether the ice meets the very tight requirements. Some geophysicists think it is unlikely. What is more, if the ice is less than 15 or 20 km thick, the thermal instabilities needed to drive the conveyor belt of convection probably cannot develop. Thick ice is necessary, though not sufficient, for convection. If the crust is convecting, it must be considerably thicker than it would be if it were transporting heat by simple conduction.

∗∗∗○ ∗

As the Galileo spacecraft approached the Jupiter system in the mid-1990s, conservative estimates of tidal heating dominated the scientific literature, not necessarily because the authors believed in the lower heating rates, but because the credibility of an ocean depended on their avoiding the appearance of extreme assumptions.

Even as the possibility of a liquid layer under the ice began to be accepted, the more extreme idea that nearly all the H_2O might be liquid remained a radical notion. Some scholars had explored the possibility that the ice layer might be thin enough for the ocean to be linked to the surface. Moreover, some of the Voyager images had shown clear evidence that plates of ice had shifted position (with much more about all that activity to come in later chapters). However, the view that Europa's crust was thick and solid was prevalent.

In itself, there is nothing wrong with approaching new data with such a paradigm in mind. Like all scientists, planetary scientists are agile thinkers, used to confronting new data and debating new ideas with open minds. The real problem is that within a large space mission there is unusual access to resources, prestige, and influence. So the stakes are unusually high and the game attracts players with complex motivations. It was no surprise to me that the initial interpretation of Galileo images was dominated by the politically powerful. The surprise was how quickly it gelled into a party line, requiring adherence to the thick-ice paradym.

The Galileo imaging team comprised a wide range of experts in various aspects of planetary science, and provided them with access to the resources and prestige of the project. Not surprisingly, the more politically skillful, aggressive, and powerful members of the scientific community had made sure they were on board.

The dozen scientists on the team were employed by various academic (or quasi-academic) institutions, with which NASA contracted for their scientific expertise. Several team members had broad power bases that extended far beyond the imaging team. They had strong influence over NASA's funding decisions, policy, publications, and public relations. Their loyalties were to themselves and their universities or research institutions. Personal motivations were varied but the prizes were scientific recognition, research money, and influence. NASA got what it needed because the team members' interests and NASA's converged where it was important. Everyone needed the mission to succeed, at least to the extent that pictures would be taken and some interesting discoveries could be reported.

But where their objectives diverged, each team member had his own agenda. The word "team" was really a misnomer, something I came to realize only gradually during my 26 years as a member. The NASA-selected team leader was Mike Belton, an astronomer at Kitt Peak National Observatory. Most people involved with the project think he did a good job: The camera got built, the pictures were taken, press conferences were held, scientific papers were published. But Belton had to do all this while playing a weak hand. To keep the team from blowing up, and to retain some semblance of control, he had to give the most powerful prima donnas pretty much what they wanted.

Those prizes included Jupiter's satellites. During the two decades between the start of the Galileo mission and the spacecraft's imaging, it was arranged that those who planned each sequence of picture-taking would get exclusive rights to do the initial interpretation of the data in those images. So if a team member could get the planning assignment for a particular moon, he could take charge of the science.

This system explained the odd events at a team meeting in Washington DC in the mid-90s. Image planning for Europa had been assigned to two notoriously powerful geologists, Ron Greeley (Arizona State University) and Jim Head (Brown University). Like Spain and Portugal dividing the new world, Head and Greeley had divided control of Europa by taking charge of alternating encounters with Europa. Head's students and assistants would plan imaging for one fly-by, and Greeley's would do the next, and so on. By picking different areas to image, they divided control of the surface and its scientific interpretation. At the DC meeting, they complained that they needed help because the job involved so much work. I had some terrific students helping me at the University of Arizona, so I volunteered to take on some of the planning work. My offer was roundly rejected. It seemed that Greeley and Head were not so interested in dividing the planning responsibilities among the team as they were in getting more money from NASA sent to their universities to pay for the job.

Those who controlled image planning won considerable dividends. They received large amounts of money for their universities, because they made the case that they needed to hire staffs of graduate students, research associates, and technicians to do the work. The overhead funds that come along with large research contracts are currency for establishing a professor's clout within the university, which translates to salary, office size, prestige, and political power. The other pay-off was that these individuals controlled the scientific results and interpretations that would be presented in the name of the Galileo project to the public and to the broad scientific community. Their initial impressions of the images of Europa, and more frequently the quick-look qualitative impressions of their graduate students and young post-doctoral assistants, were preordained to become the canonical interpretation of the character of Europa, widely accepted as established fact.

Yet given how the canonical view developed, it should not be surprising if it is wrong. It was based on initial qualitative impressions, because of contractual obligations to make quick definitive pronouncements. It was controlled by a small subset of the team, mostly geologists whose prior experience was with solid, rocky planets. The interpretative work was delegated to their inexperienced students. Other team members, with the complementary expertise necessary to do the job right, were locked out. All of this resulted in inexcusable errors in research methods and results.

***○ *

Even at its best, the "scientific method" as actually practiced is much messier than the idealized list of steps that schoolchildren are required to follow for their science-fair projects: choose a topic; form a hypothesis; design an experiment; record your data; list your conclusions. It never works out that way. With multiple researchers competing in the same arena and unanticipated screw-ups at every turn, things proceed in a much less organized way.

Moreover there is considerable variation in the practice of science from one field to another. The methods and habits of mind developed for geology are based on centuries of experience with detailed observations of the surface of the Earth. But information about the surface of Europa came from images taken of an alien planet using a telescope flying through space. We can learn a great deal from such pictures, but, if the traditional techniques of geology are used without attention to the unusual subject and novel sources of information, they can lead to trouble, and they did.

Consider the use of qualitative analogies, a standard technique in geology, where new discoveries are interpreted in terms of experience with similar features or structures. This approach has served the field well in exploring and understanding the Earth. But there is a danger in unduly applying terrestrial experience to a planet that may be completely different. The initial considerations of Europa were based on comparisons with the most similar types of familiar geological features on Earth. Inevitably then, the surface was interpreted in terms of the types of processes that operate on a solid planet. The role of an underlying ocean would not weigh heavily in such thinking.

Another problem has been that geological tradition does not address the crucial issues regarding how data about Europa were obtained. In geology, there is usually a glut of data. Researchers walk, crawl and climb all over their subject, chipping it with hammers, magnifying it under microscopes, even tasting it. In order to reconstruct and interpret the processes that produced what is seen, traditional geology needs to cope with the problem of too much data. The issue there is how to recognize underlying patterns and generalities, and how to discriminate between those seemingly minor details that are critical constraints on the big picture and those that are simply local anomalies of no broad significance.

However, with Galileo we have too little data, not too much. The problems with this kind of data are better known to astronomers than geologists: The remote sensing involved, the dependence on image data, the sparseness of the data, and the varying circumstances under which they were obtained all produce observational selection biases. This "bias" refers to unavoidable circumstances of the observations, not a failure of the scientific process. For astronomers, quantitative corrections for such types of biases are a standard part of data analysis. Bias corrections are understood in astronomy to be essential before any interpretation is possible.

Unfortunately, the initial interpretations by the dominating geologists of the Galileo project, or their students, did not take bias effects into account. In classifying features, differences in lighting and image resolution were confused with actual characteristics on the surface. The ease of recognizing certain types of features during a quick look was taken as a measure of their greater physical significance compared with features that were just as real, but harder to see. Generalizations were made on the basis of appearances in special selected locations. Anecdotal impressions became reported facts.

Ultimately, in the rush to publicize and publish the initial results of Galileo imaging, several factors came together, driving Europa's geology to be described in terms of thick ice over a largely irrelevant ocean:

- Various pre-Galileo estimates that an ocean, if any, would be covered by thick ice;
- A dependence on forced analogies with familiar features and processes on the solid Earth;
- An insufficiently quantitative analysis of the data.
- Tight control of the quick-look interpretation of Galileo data, which limited interdisciplinary consideration so that only a narrow perspective could dominate.

Despite their shortcomings, these qualitative early impressions were prominently published and widely presented at scientific conferences by the chosen few, each wearing the mantle of authority of this major mission. Those graduate students who had been assigned the image-sequence planning were anointed as the experts on Europa. Magazines for science professionals, especially *Science* and *Nature*, fast-tracked publication of the preliminary interpretations. The same story was put out by NASA's publicity machine. The party line was intensified by active promotion and by frequent repetition and citation. Those interpretations, though based on only preliminary first impressions, were quickly accepted as canonical fact.

Rather than being fast-tracked into the canon, such quick-look results should more appropriately have been treated with extra skepticism. This rule holds especially for Europa, where images show types of terrain that may reflect processes very different from familiar bodies, where data are complex and require quantitative assessment of observational selection biases, and where meaningful interpretation requires quantitative theoretical study. None of these issues were understood yet at the time of the early reports of the Galileo imaging team. Nevertheless, by early in the Galileo spacecraft's seven years in the Jupiter system, the isolated-ocean interpretation had become the canonical model of Europa.

Canonical doctrine has been known to stand in the way of scientific progress before. Resisting the canon has usually involved costly sacrifice by those who follow the scientific evidence across the boundaries of acceptable belief. In fact Jupiter's moon Europa played a central role in the case that blasted open the door to modern scientific inquiry during the European Renaissance.

For over a thousand years, the Ptolemaic theory of celestial motions had been a very successful model of the heavens: Every moving body in the sky was deemed to be in orbit around the Earth. These orbits were approximately circular, but they also involved epicycles, *ad hoc* corrections in which the bodies followed small circles around points that orbited the Earth. With these epicycles (a concept that is still useful in modern celestial mechanics), this ancient model was made to fit the observational data very well, and was an effective tool for predicting the movements of stars, planets, and the Moon across the sky.

That theoretical model also had a huge amount of political traction. Envisioning the Earth at the center of everything was appealing to anyone contemplating his place in the universe. It certainly must have seemed correct. After all, the ground appears to be quite still, with the stars and planets whizzing around the sky every day. And more important, having the Earth at the center was perfectly consistent with a literal reading of the biblical description of the relationship of the heavens to Earth. It is no surprise that this world view was considered obligatory by the western religious power structure.

The alternative model developed by Copernicus in the early sixteenth century did not seem to have much going for it. For one thing, it was not any more accurate than the Ptolemaic paradigm. Having the planets orbit the Sun, rather than the Earth, did make things simpler: fewer *ad hoc* epicycles were needed than in the Earth-centered model. But the intellectual community at the time had not yet decided whether simpler was better, although that broad question had been a philosophical issue for some time.

The merits of simplicity were overshadowed by a much bigger problem. Just as most modern U.S. planetary scientists get their research funding from NASA, and only if they stay close enough to the mainstream, Copernicus was funded by the Church, the same institution that promoted the Ptolemaic paradigm. No wonder that he was reluctant to stick his neck out publicly and publish his results. Of course in those days his neck was literally in jeopardy. Nowadays, just the job and the funding would be cut.

That kind of career damage had befallen William of Ockham in the early fourteenth century. Ockham was kicked out of Oxford for his radical philosophical ideas, and was at odds with the papal court for most of his career. Ockham's most lasting and influential philosophical contribution is known as "Ockham's Razor". The idea is that, of all possible theoretical models that could be constructed to fit observations, the simplest one, the one that shaves off non-essential details, is best. It is the one most likely to describe the essential underlying mechanism. Clearly, there is no guarantee that a simple model will be more correct than a complicated one. Sometimes a natural system may be so complicated that a too-simple model may miss what is really going on. But throughout the scientific age, experience has shown the power of Ockham's principle. Ockham's Razor underlies most theoretical modeling in modern science.

The medieval church hierarchy in the fourteenth century was prescient in recognizing the threat of Ockham's ideas to canonical doctrine. Some 200 years later, Ockham's principle gave value to the strongest asset of the Copernican model: the fact that it was elegantly simple compared with the Ptolemaic model.

In modern physics and astronomy we take Ockham's Razor for granted, so the strength of Copernicus's elegant model seems obvious. But during the sixteenth century acceptance of the Copernican model was slow. A widely accepted older paradigm seemed adequate and a powerful establishment opposed change. Then, as now, being right did not necessarily count for much and was often a liability.

The observations that finally turned the tide of scientific opinion in favor of Copernicus surprisingly had nothing to do with the controversial orbits of planets around the Sun. They involved instead celestial bodies that were unknown to Copernicus. The four largest moons of Jupiter were discovered by the great astronomer Galileo Galilei almost 70 years after the death of

Copernicus, and almost 400 years before the Galileo spacecraft imaged those same satellites. Galileo named these bodies in honor of his funding agency, the Medici family, but that was not enough to keep him out of trouble with the Church.

If these bodies were not part of Copernicus's model of planetary motion, how could the observations support the model? According to Galileo's seventeenth-century imaging data, the satellites were clearly and obviously orbiting Jupiter. He had shown by direct observation that not all astronomical bodies orbit the Earth. He had *not* shown directly that Copernicus's model was correct, but he did provide collateral evidence that it could be. That contribution was enough to change the prevailing mind-set and allow people to think about Copernicus in a different way. The canonical authorities were not pleased.

We now know these four moons as the Galilean satellites Io, Europa, Ganymede, and Callisto. Once they were discovered, the Earth-centered canonical paradigm was overturned and the door was opened to acceptance of the Copernican world view. Beyond that, the Copernican description of planetary motions led science much further. It formed the observational basis for the development of Newtonian physics. Most profoundly, this demonstration of how simplicity can bring deep understanding of physical principles gave legitimacy to Ockham's principle. It formed a template for the interrelationship between observation and theory for all of modern science.

Nevertheless, the same forces that resisted the Copernican revolution are still with us. Canonical models are challenging to shake, partly because they are usually based on substantial bodies of evidence, acquired over a long time by many researchers. Moreover, inertia always favors an entrenched paradigm. And the politically powerful, by definition, have a strong stake in the establishment position. As a result there is a strong disincentive to resist the canon, especially for the young and untenured. For all of these reasons, like all canonical beliefs, the thick-ice, isolated-ocean model of Europa has advocates among the powerful and followers among the weak. They could prove to be correct, but the mounting evidence suggests otherwise.

The images of Europa from the twentieth-century Galileo, a robotic spacecraft, were also viewed in the context of their times. There might have been an ocean, but the idea that it might be very near the surface was a radical notion. The rushed geological pronouncements were quickly made public, under an assumption that the ice was so thick that the ocean had little to do with what we see at the surface. The canon quickly solidified.

The organizational structure and policies of NASA's space program and the Galileo mission in particular also worked against innovative thinking. Galileo was an extremely expensive project, so "failure was not an option". The appearance of continual success was essential. It was not likely that Congress or the public would pay for a second try. As a result, there was a strong incentive to do things as they had been done before. Unfortunately, as I was to learn, this culture of resisting new ideas permeated all aspects of the mission.

I certainly did not know much about how these things work when I was selected in 1977 to be a member of the "Solid-State Imaging" team. "Solid-State Imaging" (SSI) is techno-speak for "digital camera". When the Galileo project got started it was known as the "Jupiter Orbiter with Probe" or JOP. Shortly afterward it was named Galileo. The obligatory three-letter acronym started appearing on memos as GLL. So I found myself a member of the GLL SSI team. There is also a very arcane difference between the Galileo Mission and the Galileo Project. The Mission refers to the overall NASA program, which was managed at NASA headquarters in Washington, D.C. The Galileo Project was the development and operational activity that was managed by the Jet Propulsion Lab (JPL) under a contract with NASA. Usually, the word Project was used as a proper noun to refer to the top project management at JPL. When JPL engineers assigned to Galileo said things like, "Project has determined that we cannot do what you asked for," they meant their bosses told them not to do it. Finally, while the GLL camera was called SSI, the camera on the Cassini mission to Saturn is called ISS, or "Imaging Sub-System." Go figure.

In October 1977, I attended the kick-off meeting of the Galileo project at JPL in Pasadena. A banner across the front of the room proclaimed "May the Force Be with Us," setting a date marker in memory: The first Star Wars movie had just come out. It was very exciting to be at what I thought was the frontier of science and technology.

I was wrong. In order to guarantee success, the project needed to stay as far back from the frontier as possible to make sure that the technology worked. When failure is not an option, neither is innovation. By the time Galileo finally launched in 1989, its technology was behind the curve: CCD cameras and computers more capable than those on board had already become consumer commodities.

To be sure, in large part the flight technology was behind the times because of extreme and unpredictable delays in the launch schedule. Galileo was originally to have been launched in 1982 from the space shuttle. Having already sold the shuttle idea to Congress, NASA needed to find some use for it, so they dismantled the infrastructure for the Titan-Centaur rocket, the old reliable work horse for interplanetary launches. At the time of the 1977 Galileo kick-off meeting, the plan was to use the shuttle, but no one knew how the space probe would get beyond Earth orbit once the shuttle got it that far. Eventually, after considerable corporate lobbying in Congress, a modified version of the Centaur upper stage was selected to fly on the shuttle along with Galileo to propel the probe onward toward Jupiter.

Selecting and developing the upper-stage rocket delayed the mission substantially. Galileo's voyage did not begin until 1985, and then it was only a road trip along Interstate 10 from southern California to Florida. As Galileo was resting at the beach in Florida at the beginning of 1986, waiting its turn for a shuttle launch, everything changed with the explosion of the shuttle Challenger.

In addition to the delays inherent in checking out and re-establishing the shuttle program, it became clear that a bomb as big as a modified Centaur rocket was not going to be allowed to ride along with humans on the shuttle. The whole strategy for getting Galileo to Jupiter had to be re-worked. So the next leg of Galileo's journey was not toward Jupiter, but instead back over the I-10 highway to California.

As Galileo waited in California again for the next several years, civilization marched on: the shuttle program was reinstated; a weaker "Inertial Upper Stage" rocket was developed to propel Galileo onward; and Galileo's once-advanced on-board technologies became dated relative to consumer commodities. Finally, Galileo rode across America a third time and was launched from Cape Canaveral in late 1989 on the shuttle Atlantis. The Inertial Upper Stage was so weak that extra encounters with Venus and the Earth were needed to give gravitational kicks to the craft, at the expense of doubling the travel time to Jupiter. The bundle of shop-worn 1970s technology reached the Jupiter system late in 1995 and didn't complete its mission until 2003, more than a quarter-century (and four more Star Wars episodes) after the kick-off meeting. Call me superstitious, but that 1977 Stars Wars banner was asking for trouble.

Even if Galileo had gotten to Jupiter on its original schedule in 1985, its technology would have been old. During the first years of the project, as I listened to the reports and discussion at endless imaging team meetings, I noticed that the selection of components generally seemed to require that they be available in commercial catalogs. While it was disappointing not to be as close to the forefront of technology as I had thought, it did make a kind of sense. Anything that flew on the spacecraft had to work, and that meant it needed a track record.

But what was disappointing and inexcusable was the way this caution pervaded all aspects of mission planning, extending even to planning for the science analysis that would be done here on planet Earth. What well-managed enterprise would specify technology purchases and functionality five years into the future on the basis of what was already available on the market?

It was not until 1980 that I began to recognize this aversion to the cutting-edge of technology, and then only because the Hunt brothers of Texas had tried to corner the market on silver. Photographic prints—that is, photos developed and reproduced on paper—are made with silver. Mission planning in the late 1970s included consideration of the production and distribution of the pictures that the spacecraft would take when it got to the Jupiter system. For budgeting picture production, planners estimated that 100,000 pictures would be taken, and each would need to printed in a few different versions, with varying contrast, brightness, sharpening, and other enhancements. All these large format pictures would need to be distributed to all the imaging team members and to various archives. You can do the arithmetic: the project would need literally millions of prints. By the time the Hunts bought up half the world's silver supply, the Galileo darkroom budget soared to $12 million.

Several things about this plan impressed me. The first was that the spacecraft was expected to be able to get us much more than the 100,000 pictures in the budget. The number of pictures would not be limited by the durability of the spacecraft, nor by the rate at which the digital images could be radioed back to Earth. Instead, unbelievably. the number was going to be limited by the staggering printing costs.

I was also impressed by the dollar budget. With 12 team members, the mission would be spending a fat round million dollars on photo printing for me. It was not obvious that this plan was how I would choose to spend my share of the resources. Finally, I could hardly imagine doing science with hundreds of thousands of large-format photographic prints. The mere physical manipulation seemed mind-boggling. I could hardly imagine organizing or analyzing them in any systematic or quantitative way.

Nevertheless the plan followed space-mission tradition, the tried-and-true way. The bits of data would be radioed back to Earth. Each image was to be composed of 640,000 measurements of brightness in the camera's field of view, arrayed in a field of 800 × 800 pixels. The brightness values were encoded as 256 levels of gray (256 levels = 8 bits = 1 byte), so each image could be encoded as 640,000 bytes or about 2/3 of a megabyte – or a few times less with data-compression schemes. A central processing facility on Earth would spend $12 million converting that information into photographic hard copies for delivery to the science teams. Once the hardcopies were printed, the original bits of data would be stored on magnetic tape, where they would be hard to access, because tape is cumbersome to manage and vulnerable to physical deterioration and data loss. Following time-honored convention, photographic hard copy would be considered the final, permanent product.

As a medium, photographic hard copy has some advantages. The printed resolution can be very high; you can quickly look over many images spread across a table; no special viewing equipment is needed; and they can be tacked to a bulletin board. However, compared with digital information, there are overwhelming disadvantages. A photographic print cannot display the full range of brightness and contrast that are contained in the original digital data. Using (and then archiving) hardcopy, rather than the original bits, means that you are throwing out a good part of your expensive and valuable information. Once hardcopy is printed, further enhancements are limited. Even worse, most photographic processing is done with automated routines that may hide the most valuable of scientific data: the unexpected.

I had watched as Voyager images were processed back in 1979. A team of skilled craftsmen from the US Geological Survey was detailed to JPL, with sharp knives and glue, to cut-and-paste photos into large mosaics. Those people were very capable and hard working, but the activity was like something from the era of the Second World War, rather than the new age of information technology.

Because it depended on printed-on-paper images, scientific interpretation was hobbled during the Voyager encounter with the satellites of Jupiter. For example, when the Voyager imaging team received the first photos of satellite Io, they were mystified by its appearance. The photos showed a strange blotchy place that did not look like any planetary surface that they had seen before (**Fig. 3.1**, left). In fact, the explanation was clear in the bits of information that had been radioed to Earth, but it was lost when the photos were printed. Fortunately, *different versions of the same images* were prepared for the optical navigation team. Those engineers were given overexposed, high-contrast versions because they needed to see the stars in the field of view of the camera in order to navigate the spacecraft (like Fig. 3.1, right). Io appeared only as a white disk, but it was this overexposure that allowed an optical navigation engineer at JPL, Linda Morabito, rather than an imaging team scientist, to discover volcanism on Io when she spotted the gigantic volcanic plumes spraying into the dark sky. The information for both versions of the image was in the original digital bits sent by Voyager, but any single print could only show part of the story.

At the time that the Galileo imaging team was planning its photo budget,

FIGURE 3.1: At left, a Voyager image of Jupiter's moon Io that initially baffled scientists. At right, the same digital data, when printed to enhance the background sky, showed a fountain shooting into the sky, and revealed Io's volcanic nature.

elsewhere computers were already being used to display and process images. The field of digital image processing was still in its infancy, but its potential was clear. The commercial sector, already active in the Silicon Valley, was developing applications poised to revolutionize image processing in all sorts of fields, especially where the money was – in biomedicine and defense. But on the frontiers of planetary science, revolutionary digital image data, radioed hundreds of millions of miles from Jupiter, were being processed using nineteenth century technology. Galileo's $12-million plan for the future was for more of the same.

I decided that I would prefer to let the Hunt brothers keep my share of the silver. I would be happy to forego any hard copy if the Galileo project would simply send me the bits of information when they arrived on Earth. All that I asked in exchange was that I might get part of the million-dollar savings from my share of the darkroom expenses to buy image-processing computer equipment. The advantages seemed obvious, at least to me. The project would save money. I could buy a great computer system, and by waiting until I needed it I could buy an even better and cheaper one. I would not have to manage a huge library of photographic hardcopy in my lab. And I would have the digital bits to look at any way I wanted to. Moreover, I would be able to do precise quantitative measurements with greater ease and precision than with paper photos. It seemed to me that this approach might work for others on the team as well.

I gave a presentation of this idea at a team meeting in Hawaii in 1980. It did not go over well. The reasons seemed incredible to me at the time, and are even more incredible in retrospect. For example, one geologist could not understand how you could make measurements on a digital image. He was unable to envision modern graphical measuring tools on a computer. He thought that my proposal would require him to lay a wooden ruler against his computer monitor. The old-timers could not imagine changing their ways. Everyone reacted negatively. As events later unfolded, I came to understand another underlying reason for the embarrassing reception: the broad distribution of resources implicit in my vision could only upset entrenched interests. My idea was roundly mocked and rejected, but I learned an important lesson: At the frontiers of planetary exploration, new ideas are not necessarily welcome.

My plan was at odds with another aspect of the Galileo mentality. The planning process had no way to accommodate the rate of change of technological advancement. The plan for 1985 had to be based on computers that could be found in a 1980 catalog. The same went for data storage media. My plan was attacked on the grounds that it would require rooms full of computer tapes to store the data at each of our home institutions. No one bought my prediction that a few laser disks would be able to hold the whole Galileo data set on a single bookshelf. Actually this was not a radical idea at all. Video disks were already on the commercial market and the CD format was in the works, but such notions were too risky for Galileo. In short, the Project was as conservative in planning technology for my office, which was to be bought years in the future, as they were in planning for the technology that was to fly into space. If I had understood the politics and mentality of the project and imaging team, I would have kept my mouth shut. Being right did not do you any good, unless you kept it to yourself. Sharing good ideas was definitely a mistake.

Of course, over the next few years as technology marched on, the project had to let go of its old ways. The whole world was adopting digital imaging; image processing could be done on home computers; and the whole Galileo data set could fit in a sun-visor CD caddy. I now have all the images of Europa on a single CD. Eventually the imaging team adopted my plan, except that somehow, as the money flowed, some team members were more equal than others.

The same resistance to change that affected use of technology also permeated the imaging team's way of doing science. For Europa the thick-ice, isolated-ocean paradigm was embraced because any other interpretation would have suggested a high-profile commitment to something new. Authorized speakers propagated a mantra of questionable facts and anecdotal evidence selected to bolster the model. The isolated-ocean model was widely disseminated as the authoritative Galileo mission dogma.

At the same time as that paradigm was becoming entrenched, a less rushed, interdisciplinary review of the data began to show that what is seen on Europa is actually quite different from much of what had been reported earlier. Some key anecdotal evidence from the earlier quick-look proved to have been premature. Observational selection effects were proving to have been significant, and quantitative corrections could be made. An interdisciplinary approach, in which the strengths of geological methodologies are enhanced by complementary ways of representing and interpreting the data, has now produced a very different big picture of the character of Europa, in which the ice is thin enough for the ocean to be linked to the surface.

At first, this permeable crust model faced considerable opposition. Like all canonical paradigms, the isolated-ocean model would be hard to unseat. Part of the resistance to change may have come from the fact that this subject was initially relegated to the realm of geology. The use of broad, underlying conceptual models, like other aspects of scientific methodology, differs among various disciplines. In astronomy, theoretical models are constructed to connect the dots between sparse data; conceptual models must be rebuilt frequently as new information becomes available. In geology, such broad conceptual models are harder to construct, because so many pesky details get in the way of the big picture.

Geologists seem to have a harder time shifting paradigms. The field of geology may never live down the story of continental drift during the first two-thirds of the twentieth century. Any layperson with a world map could see that the Americas had broken away from Europe and Africa. But geologists knew too many details that did not seem to fit the simple story. Acceptance of the global processes of plate tectonics and continental drift was resisted for decades. But once those broad principles were accepted, the details fell into place.

A cautious resistance to paradigm shifts is reasonable when a model has been serving well. But the isolated-ocean model for Europa had become the canonical paradigm for all the wrong reasons. Now, however, a very different interpretation of Europa is emerging, in which intimate linkages with the ocean have continually reshaped and replaced Europa's surface. This result is based on quantitative investigations of the tidal processes that drive activity on Europa and quantitative analyses of observations, especially correcting for observational selection effects that may have skewed initial impressions of the surface character of the satellite. As we shall see, one of the great strengths of the new permeable crust model is that nearly all of the major observed characteristics of Europa's surface can be readily explained with a single assumption, that the ice is thin enough for cracks and occasional melting to expose the ocean.

In what follows, I describe the various lines of evidence and show how they have led to this emerging picture of Europa's history and physical structure. While we have seen the folly of accepting any scientific model as dogma, the evidence for this broad picture is compelling. If we are correct, the physical setting on Europa, with its ocean linked to the surface, provides potentially hospitable environmental niches that meet the requirements for survival, spread, and evolution of life. If there is life on Europa, the biosphere extends upward from the ocean to within centimeters of the surface and frequently breaks through. And everything is driven by tides.

Planetary Stretch

On Earth, twice each day, tides wash up over beaches, flow in and out of estuaries, and fill and empty harbors. At low tide in New Brunswick, Canada, water drips onto the beach from wet seaweed on cliffs 16 meters above. Yet there are no discernible tides in the Mediterranean, and sailors crossing the oceans see no tidal effects at all. From human experience, we tend to think of tides as a local phenomenon, where twice each day the ocean rises and falls against the land. But on Europa there is no land. With an ocean 150 km deep covering the entire globe, how can there be tides?

From a whole-body point of view, tides are a very different thing. The tidal effects at a particular beach or harbor are simply a detail of the global picture. As diverse as the tides seem from a local or regional perspective, we can understand them by thinking on a global scale.

The self-gravity of a planet tends to make it spherical. Each bit of mass pulls on each other bit, with their cumulative gravity pulling the material into a symmetrical ball. The strength of the material can support some deviations from the regular global shape dictated by gravity, but not much. And eventually shaking and erosion allows gravity to pull mountains down toward its center, continually working to keep the planet round.

Tides come into the story when we consider that no planet is alone. Besides self-gravity, any neighboring body also exerts forces on each bit of the planet's mass. Our Moon acts on the Earth, for example. The force of gravity depends strongly on distance, so the Moon pulls more strongly on the part of the Earth nearest to it, less strongly on the center of the Earth, and even less strongly on the mass in the Earth on the side farthest away. The Moon stretches the globe of the Earth (Fig. 4.1), elongating it into an oval (more precisely, the 3D version of an oval, an ellipsoid). What matters for this stretching is not how hard gravity pulls, but rather how much harder it pulls on one part of the Earth compared with another. The amount of stretching is nearly the same on each side, so the two tidal bulges are equally high.

The key point here is that you should stop thinking of a tide as just a local

FIGURE 4.1: The Moon pulls harder (long arrow) on the nearest side of the Earth, and less (short arrow) on the farthest side, so the whole Earth gets stretched. The Earth does the same thing to the Moon, but that stretching is not shown here. This sketch is schematic, and not to scale.

thing that happens at your boat dock or favorite beach. A tide is a global effect. It stretches a moon or a planet (in this example, the Earth) into an oval shape, elongated in line with the other body that causes the tide. Nothing about tides requires an ocean. The whole planet stretches, no matter what it is made of. The rocky body of the Earth stretches under the tidal pull of the Moon, although the strength of the rock resists the pull. The watery envelope around the Earth elongates more, but only because it is weaker than the rock.

Things get more complicated in the real universe. Every other body, not just the Moon, is also pulling and stretching the Earth. Fortunately for the folks who have to calculate these things, only one other body makes a significant contribution, and that is the Sun. It acts to stretch the Earth toward itself, so this stretching is in a different direction and different amount than the tidal stretching due to the Moon. Even though the Sun is some 30 million times more massive than the Moon, it also is 300 times farther away. By coincidence these effects roughly balance, so it and the Moon each stretch the Earth by a similar amount. So at any instant, the globe of the Earth is being stretched in two different directions by similar amounts. Each of those two tide raisers, the Sun and the Moon, tries to mold the Earth into an ellipsoid elongated toward itself. It turns out that, when you do the geometry, the total of those two distortions is also an ellipsoid. But because the Sun's position relative to the Earth and Moon is always changing, the magnitude and direction of the elongation periodically changes.

The shape of the Earth continually responds. On average the surface of the ocean layer is pulled about 3 meters upward, at the ends of the stretched ellipsoid. Each of Earth's spherical layers—its metallic core, the rocky mantle and crust, the atmosphere (as well as the thin water layer)—also stretch into ellipsoids, all conforming to the combined gravitational pulls of the Sun, the Moon, and the self-gravity of the Earth.

But the Sun and the Moon and the Earth do not stay still. The Moon orbits the Earth each month; the Earth orbits around the Sun each year; and the Earth spins around each day. The direction and magnitudes of the tides keep changing. The materials of the Earth, the nickel-iron of the core, the rock, the water, and the air, are all subjected to ever-changing tidal forces that continually deform and re-mold them.

The planet simply cannot keep up. The rock that makes up the bulk of the Earth cannot stretch or flow fast enough to keep the shape lined up with the tidal forces. The alignment is always a bit behind where it should be. Moreover, the stretching of the rock never gets nearly as elongated as it would be if it had time to respond.

Only the fluid spheres of Earth—the ocean and the atmosphere—can respond in a reasonably timely way. But the ocean layer is thin, only a few kilometers at its deepest. In order to change the outer shape of the planet, the water flows to fill in each end of the stretched ellipsoid with an extra meter or so of fluid. The oceans must keep rushing in changing directions in response to the tidal force as it moves around. Pesky continents get in the way, even though they cover only a small fraction of the surface. The water flows and sloshes in all sorts of complex ways as it rushes, trying to respond to the changing tidal force. The flow gets trapped in places, funneled into random coastline configurations like Canada's Bay of Fundy. The water moves especially fast as it passes through the Bering straits and across the shallow Bering Sea. The Mediterranean Sea is so land-locked, that there is not much change at all; relatively little water can squeeze past Gibraltar.

The local effects vary tremendously from place to place, but the tide is a global-scale phenomenon: the continual reshaping of the Earth as its fluid outer layer accommodates to the tidal pull of the Sun and Moon.

On Europa tides are much simpler, and much larger. Only one tide-raiser dominates, and that is Jupiter. It is so massive (30,000 times the mass of the Moon) and so close to Europa (only about 1.5 times as far as the Moon is from Earth), that its tide stretches the 3300-km-wide globe of Europa by about 1 kilometer. Compare that value with the 13000-km-wide Earth being stretched by only a few meters.

Europa's tide is huge, but it changes much more slowly than the Earth's. Compared with the Earth's oceans, Europa's water layer has an easier time keeping up with the global shape dictated by the tidal forces. It helps that there are no continents or islands or shoals poking up from the rock below, so the water can readily flow to conform to the ellipsoidal global shape. And the ice on the surface is too thin to resist that changing shape of the water layer. Below the ocean, the rocky mantle and iron core of Europa also have time to conform by slow viscous flow to the stretching effect of Jupiter.

The reason this tide is nearly constant is that Europa keeps one face toward Jupiter. In other words, as Europa orbits around Jupiter once every 80 hours, it rotates at a matching rate. Try this at home: Walk once around a friend keeping one shoulder toward him or her. Note that in the process your body will spin around exactly once relative to your surroundings. Such synchronous rotation is common among satellites in the solar system. Our Moon of course is the classic example. We always see the same face of the Moon from Earth.

Remember, the tidal force on any moon due to its planet tends to stretch out the moon, bulging it on the planet-facing side and on the opposite side. For any moon in synchronous rotation (e.g., our Moon or Jupiter's Europa), those bulges will always be in the same places on the body. The material (rock, water, or whatever) does not need to constantly shift around trying to catch up with the changing direction of the tidal force. That is why Europa's enormous tide, kilometers high, is nearly fixed in place on the body. Europa's body and ocean are stretched close to the shape dictated by Jupiter, and they do not have to shape-shift very much.

In fact, if Europa were on a perfectly circular orbit and spinning perfectly synchronously, there would be absolutely no variation in this tide. The amplitude would not change, because the distance from Jupiter would be constant. The orientation of the elongation relative to the body of Europa would also be fixed, because with synchronous rotation the body retains a fixed orientation relative to the direction of Jupiter. As huge as this tide is, other than establishing the shape of Europa, it would have no interesting effects.

And, indeed, most moons are that way. In fact, the reason they are that way has to do with tides. Most moons (including Europa) probably started off spinning in fairly random directions and moving on eccentric orbits, depending on the last big impacts that happened during their formation period. With non-synchronous rotation, the material of the moon would continually play the catch-up game, and the tidal elongation would never quite be lined up with the direction of the planet. The asymmetry of the shape would allow the planet's gravity to exert a torque (a twisting force) that would eventually force the moon to rotate synchronously. Similarly, it also would tend to make the orbit circular. So tides are sort of suicidal, or self-canceling. They make spins synchronous and they make orbits circular. At that point the tidal shape of the body becomes fixed and constant—boring—exerting no tidal effects at all.

The good news from Europa was that the tide is not exactly constant, so some very interesting things can happen. This was good news for me because we celestial mechanicians love to work out these complicated processes. What's more important, though, is that Europa's non-boring tides would be very good news for any organisms on Europa, because the things that tides do might make Europa habitable.

So why didn't Europa's tides turn themselves off? Why did the usual boring moon tide scenario not occur there? And why did it take so long for people to figure it out? The plot thickens further.

The first evidence for something much more exciting than usual came from Galileo's data—that is, from the great astronomer's observations of the moons of Jupiter in the early 1600s. Galileo made careful plots of the movements of the satellites that he saw through his telescope. The orbital periods were quite well defined. The periods of the inner three are approximately the following: 42 1/2 hours for Io, 85 hours for Europa, and 170 hours for Ganymede.

The ratios of these periods are striking. From these numbers, going back to the original observations by Galileo, it appears that they are locked into an exact ratio of 1:2:4. In the time it takes Ganymede to go around Jupiter once, Europa goes around exactly twice and Io goes around four times. This ratio also means that the location in the orbit where Io overtakes Europa, where they are both in line with Jupiter ("conjunction"), is always the same. After a conjunction, Io advances ahead of Europa, and next catches up with Europa at the same location, after Europa has made one whole orbit and Io has gone around twice. The conjunction of Europa with Ganymede also occurs at a fixed location, because of the 1:2 ratio of their periods. Moreover, Galileo's plots of the orbital motion also showed that the conjunction of Europa with Io always occurs at exactly the opposite side of Jupiter from the conjunction of Europa with Ganymede.

The scientific world at the time of Galileo found it sufficiently traumatic to deal with the simple issue that the satellites seemed to be orbiting Jupiter and not the Earth, in violation of canonical wisdom. It took almost two centuries before the strange and amazing ratios among the orbital periods were understood. The explanation was part of the great body of research carried out by the French mathematician Pierre-Simon de Laplace, and published early in the nineteenth century. Laplace showed that the periodic repetitions of the conjunctions of the Galilean satellites enhanced their mutual gravitational effects.

Although each satellite's orbit was predominantly governed by Jupiter's gravity, the satellites' effects on one another can perturb the orbits in various ways. Whatever effects they have on one another are reinforced by the way their mutual configurations repeat periodically, every few days, because of the remarkable ratios among their orbital periods. This enhancement, due to the periodic repetition, is a *resonance.*

The resonance has several effects. For one thing, it is self-preserving, by maintaining the whole-number ratios of the orbital periods. Most important is that this resonance maintains the eccentricities of the orbits. Even though tides tend to circularize orbits, the resonance among the satellites keeps them on eccentric (elliptical) paths.

Remarkably, though, more than 150 years after Laplace, as astronomical interest in the planets and their moons grew during the twentieth century, a misunderstanding led to a widespread belief that the orbits of the Galilean satellites were circular. In nearly every textbook and authoritative tabulation of orbital parameters, the eccentricities of the Galilean satellites were listed as precisely zero. A perfectly circular orbit has an eccentricity value of 0, while the more "stretched" or elliptical it is, the higher the eccentricity value.

The misleading table entries had propagated from a peculiar tradition in celestial mechanics. For satellites in resonance, eccentricity values can oscillate. What was the poor table writer to do? The actual eccentricity at any time would be something different from what the table showed, and ever-changing. Given limits on available columns and given the need to keep tables simple, often only the amount of the periodic variation is tabulated. It happens that for Io and Europa, resonance creates substantial eccentricity values, but the periodic variation is near 0. So, for many decades, astronomers and students were informed by the authoritative reference literature that the eccentricities of these satellites were 0. The true eccentric motion remained unappreciated. (One reason that no one was surprised to see the value 0 listed is that tides were expected, in general, to circularize orbits, especially if you did not fully appreciate the resonance.)

In the mid-to-late 1970s, I wrote a few articles about resonances among satellites, and gave several presentations about them at conferences. A few weeks before Voyager reached the Jupiter system, I got a phone call from my colleague Stan Peale, a Professor of Physics at the University of California in Santa Barbara. Stan is a few years older than I, and I had learned a great deal about tides and celestial mechanics from studying his work. This time Stan had questions for me. He had heard me speak about the Laplace resonance, and he wanted to confirm that, contrary to the common tabulations, the eccentricity of Io was 0.0041 and the eccentricity of Europa was 0.01. Had he understood me correctly when I had said to disregard the tabulated values of zero? I told him that was correct. "Do you know what this means?" he asked.

"No, what?", I cluelessly replied.

"It means that there must be an incredibly high rate of tidal heating in Io and probably in Europa as well." I became the first of many celestial mechanicians to slap palm to forehead: Why didn't I think of that!

***O *

As I described above, if a satellite's orbit is circular and its spin is synchronous, the tidal bulges become fixed. Even if the bulges are big, they will have no interesting geophysical effects because they are fixed and unchanging. There is no friction, and so no heating. But, if a resonance keeps the orbit eccentric, the tidal effect increases and decreases during each orbit, as the satellite's distance from the planet changes. The body continually changes shape, so there is friction and thus heating. And, if the tide happens to be raised by a nearby planet that's the largest in the solar system, a great deal of heat must be produced.

A few days before Voyager reached Jupiter in 1979, the magazine *Science* published the article by Peale *et al.* that predicted something dramatic would be found at Io and possibly at the other Galilean satellites. The rest is history: Voyager found Io to be actively spewing volcanic plumes into space, and the surfaces of the other resonant satellites, Europa and Ganymede, appeared far more active than frozen ice balls in space had any right to be.

Years later, what had been learned about the Galilean satellites from Voyager was reviewed in a reference book, *The Satellites of Jupiter.* In my chapter in that book, I covered the dynamics of the Laplace resonance and described the values and variations of the eccentricities. I included a table that showed the value of the forced eccentricity due to the resonance, 0.0101, and the tiny variation of 0.00009. (The variation is so small that the actual eccentricity is always near 0.01.) Nevertheless, the opening chapter of that same book contained the same old incorrect table summarizing the orbital parameters of Jupiter's satellites: Europa's eccentricity is listed as 0.000. Once ideas get established they are hard to change.

Ignore what you may read elsewhere to the contrary. Europa's orbit is not circular. Its eccentricity is about 0.01. Everything that is interesting about Europa follows from the fact that the eccentricity is not 0.

Because Europa's orbit is eccentric, tides on its body change over the course of each 85-hour orbital period. **Fig. 4.2** represents schematically how the orbital eccentricity drives the tidal variation. The magnitude of the tide depends strongly on the distance from Jupiter. The tide-raising gravitational effect is strongest when Europa reaches its minimum distance from Jupiter (at *perijove*, which means closest to Jupiter) and weakest half an orbit later at

FIGURE 4.2: The change in Europa's tidal figure between perijove (closest point in the orbit to Jupiter) and apojove (farthest point). The body of Europa must continually remold in order to conform to the shape driven by the tide, with the strongest tide, and hence greatest elongation of the body, at pericenter.

Europa rotates synchronously, so that a flagpole planted on the Jupiter-facing side would point toward Jupiter throughout each orbit.

Note: This schematic shows the general idea but it is not drawn to scale: Jupiter's diameter is actually 50 times Europa's; Europa's orbital diameter is about 10 times Jupiter's and the orbit is more nearly circular with an eccentricity of only 1%; the tidal elongation of Europa averages only 0.03% of Europa's diameter and varies by only 3% of that amount between apojove to perijove. If drawn to scale, nothing of interest would have been visible. (Illustration by Barbara Aulicino/*American Scientist,* based on my sketch.)

its maximum distance (at *apojove*). Bear in mind the true proportions of the geometry: Europa's radius is about 1560 km and it averages a distance of 671,000 km from Jupiter, whose radius is about 71,500 km. The average height of each tidal bulge on Europa is about 1/2 km. Europa's orbital eccentricity is enough that Europa's distance from Jupiter varies by about 1%. That small variation in distance means that the orbit is nearly circular, but the variation is enough to drive dramatic effects on Europa.

The tide does not only change in magnitude. Over each orbit the direction of the elongation relative to the body of Europa changes as well. The explanation follows from something Johannes Kepler discovered in the early 17th century about the orbits of planets, something that applies to Europa as well: Europa must go extra fast in its orbit around perijove and extra slow around apojove. But Europa's rotation (its spin rate) is essentially constant over the whole orbit. Think about what that means for a flagpole pointing to Jupiter at pericenter. With the extra orbital speed there, the rotation of

Europa is too slow to keep the flagpole pointing exactly toward Jupiter. Later, as Europa approaches apojove, its orbital motion slows down compared with the rotation (which stays constant), and the flagpole catches up again with the direction of Jupiter. The tide always wants to be aligned with the direction of Jupiter, but that direction shifts back and forth relative to the flagpole, and thus relative to the body of the satellite.

So the orientation of the tide relative to the body of Europa varies periodically during each orbit, just as the magnitude varies. The body, including its rocky interior and thin ice crust, as well as its thick liquid water layer, is continually re-molded in response to these changes in magnitude and direction of the tide. (Understand that the change in direction of the bulges does not represent rotation of the body, but rather a re-molding of the moon's shape.) Every 85 hours at the imaginary flagpole (and simultaneously at the opposite side of the satellite) the height of the surface rises and falls by 30 m relative to the average tidally elongated figure of Europa. And between the rising and falling, the high point of the bulge moves back and forth (by about 100 m) from one side of the flagpole to the other.

At least that is how much the shape of Europa would change if the materials could respond quickly to the tidal force. Most calculations show that the thick liquid layer that dominates the outer 150 km of Europa would allow a rapid response. So as long as Europa's water is largely liquid, our estimate of the amount of tidal variation is probably accurate. But if the water were nearly all frozen, the magnitude of the tide would be much smaller.

If spacecraft ever return to Europa, they may be equipped with altimeters capable of measuring the height of the tide. In principle, these measurements could detect the thickness of the ice. Unfortunately, other uncertainties about the internal structure may swamp out the effect of ice thickness. Measurements of tides will tell us whether there is an ocean, but only with very precise measurements and improved interior models will there be any hope of using this approach to determine the thickness of the ice.

Because the length of a day on Europa (which depends on its rotation period) is very close to the period of an orbit, we have come to call the tidal variation the *diurnal tide*. I have an uneasy feeling that I may have introduced that term to Europa science. If so, I regret it because the periodic changes that I just described depend on the orbital motion more than on the spin. But the name stuck because the period is essentially the length of a day on Europa, and only three-and-a-half Earth days. Whatever you call it, these periodic workings are responsible for the unique character of Europa, providing frictional heat for melting the water, and stressing and cracking the icy crust.

***O *

Were it not for the eccentricity-driven diurnal tides, Europa would be as inactive as Callisto, the farthest out from Jupiter of the four Galilean satellites. Callisto, not part of an orbital resonance and thus with no diurnal tides, is simply a heavily cratered target for bombardment by every type of debris in the solar system. It is not being changed from within. The three resonant satellites (Io, Europa, and Ganymede) all have much more interesting geology and geophysics because of the various effects of the tides.

Tides affect the satellites in four major ways:

First, the continual remolding of the figure of the body entails friction and so generates heat. We usually think of frictional heating as something that happens when two objects are rubbed together. Here we are talking about internal friction, energy dissipated when work is done distorting the material. Bend a paper clip back and forth and it gets hot. The same thing happens when the materials of Europa, iron, rock, and ice, undergo the repetitive distortion by diurnal tides. Europa gets heated from the inside out.

Second, the change in shape stresses the material. We haven't got a clue about the effects of this stress in the deep interior, but we certainly see its effects on the ice crust. Imagine trying to change the shape of an egg. The fluid below the surface would not be affected much, but the shell would certainly crack. The same thing happens on Europa. Hour by hour its shape is changed. The ocean water readily accommodates, but the ice shell must crack, and then each crack continually gets worked. Europa's surface undergoes a variety of tectonic processes evident on its surface.

Third, as the tidal force changes, the response of the material of the satellite lags behind a tiny bit. So the elongation of the satellite is not exactly lined up with the tide-raiser, Jupiter. Jupiter's gravity pulls on this ever-so-slightly asymmetrical mass distribution in a way that can apply a torque (a twisting force) on a satellite's rotation. Europa may not rotate synchronously after all.

Fourth, those same gravitational torques, as well as torques involving tides raised on Jupiter by the satellites, have probably caused long-term variation in the orbits of the satellites. The Laplace orbital resonance may have changed over time. And that means Europa's orbital eccentricity may have been different in the past.

The heating and stress generated by tides have direct effects on the observable surfaces of the satellites. The stress corresponds to observable tectonics like cracks, and heat corresponds to observable thermal features like chaos. The changes in rotation and orbits modify the tides and thus indirectly affect what we see on the surface. All of these processes are interdependent, as summarized in **Fig. 4.3.**

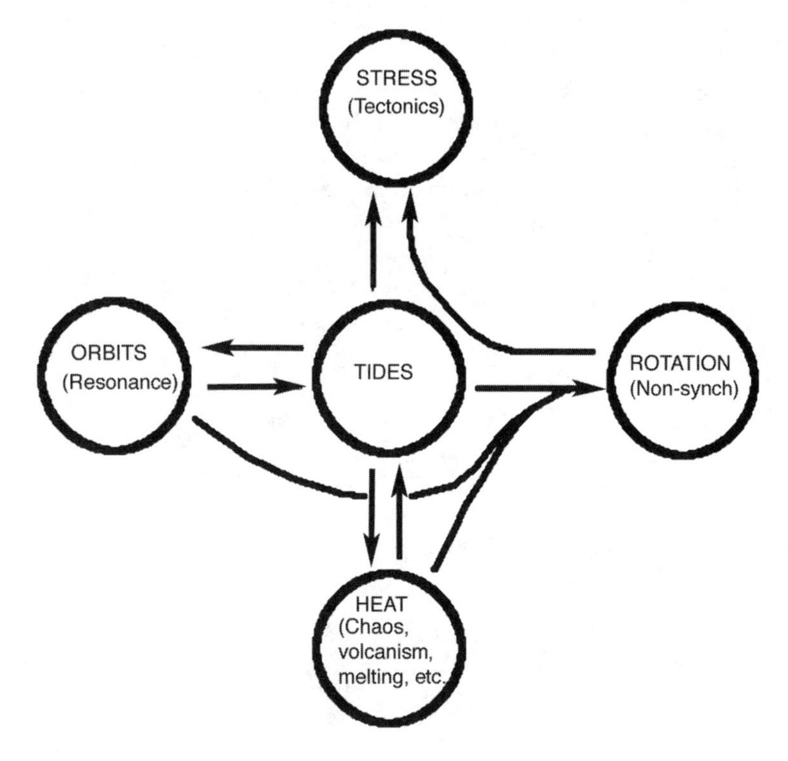

FIGURE 4.3: Tides lie at the hub of the interrelated processes that govern Europa's geology and geophysics. Arrows show where one type of phenomenon affects another:

(1) Tidal distortion stresses the surface cracking it and driving tectonics.
(2) Tides change the satellites' orbits around Jupiter. In return, the orbits (via resonance and eccentricity) drive the tides.
(3) Tidal friction heats the satellites. In return, the heating can modify the material and hence the magnitude of the tidal stretching. The height of a tide depends strongly on whether part of the interior has been melted.
(4) Tides may drive non-synchronous rotation, but only (see the lower curved lines) if the satellite is well heated and in an eccentric orbit; otherwise tides alone would result in synchronous rotation, as they did for the Earth's Moon. Non-synchronous rotation in turn modifies tides in a way that may profoundly affect stresses on the surface (upper curved line), because the tidal bulges would move continually around the equator, as they do on Earth.

These effects exist on all of the Galilean satellites, but they do not necessarily manifest themselves to the same extent or in the same way. On Io, because it is closest to Jupiter, the tidal heating is so great that volcanism swamps out and hides most large-scale tectonic effects. It is not obvious whether any of the tectonic features observed there are related to tides. At the opposite extreme, far from Jupiter, poor pathetic Callisto has no significant tides because it is not part of the orbital resonance. Ganymede shows little indication of heating because it is, of the three resonant satellites, the farthest from Jupiter. (Nonetheless, a substantial portion of Ganymede's surface is dominated by tectonics.) Europa lies in the "sweet spot" between Io and Ganymede, where there is considerable tidal heating, but not so much that thermal effects hide the tide-dominated tectonics.

A Closer Look at Tidal Effects

Interpreting Europa depends on understanding the four effects of the tides: the effects on heat, stress, rotation, and the orbit.

In considering the *heat* produced by tidal forces, it may be best to start with an overview of Jupiter's four large moons, moving from the closest to the most distant: Io, Europa, Ganymede, and Callisto.

As the figure of a satellite is distorted periodically due to diurnal tides, friction generates heat. When Stan Peale and his colleagues in 1979 put that old idea together with the long-known orbital behavior of the Galilean satellites, they realized how much heat might be generated within those bodies. For Io, closest to Jupiter of the four Galileans, the calculations led to their legendary prediction that Voyager images would show major thermal effects. Only a few days later, huge amounts of on-going volcanism were discovered there. The Voyager images of Europa were not so dramatic, but it was clear that this satellite, too, had been geologically active until recently, and there was no reason to believe that the activity had stopped. The contrast between Europa's appearance and that of dead, cratered objects was obvious.

Also obvious was the clear trend among the Galilean satellites correlating the geological activity with distance from Jupiter. The closer a satellite is to Jupiter, the fewer impact craters and the more the geology is dominated by tidally driven processes. Furthest out and not part of the eccentricity-driving Laplace resonance, Callisto was the archetype of a dead bombarded planet. Closer to Jupiter, and with a resonance-driven eccentricity, Ganymede displayed large Callisto-like areas, but other broad regions showing tectonics and resurfacing. Closer still was the completely resurfaced Europa, and closest to Jupiter was the spewing Io.

Given the great activity at Io, it was natural to calculate the likely effects of tidal heating on Europa, the next satellite in the resonance. The relevant properties of Europa could only be guessed at, especially the parameter Q, which is supposed to represent and characterize all of the diverse frictional effects within the satellite. Q is a universal parameter used in engineering and physics to represent the dissipation of energy into heat in any oscillating system. For example, your car's suspension has springs (oscillators) that are damped by shock absorbers, which dissipate the energy from bumps into heat.

To make a credible case that something weird might be going on within Europa, the theorists needed to adopt very conservative assumptions about the value of Q, and everything else that could affect the rate of tidal heating. In other words, they needed to show that interesting things could result with even the minimum likely amount of heat. They found that even with such minimal-heating models, tidal friction could keep most of the H_2O melted.

So the heat was enough to maintain a global liquid water ocean, a major discovery in itself. But theorists always try to push the implications of their work as far as they can. They realized that, without a global ocean, the tidal stretching of the satellite would be small, so the crust would not be stretched much, and there would be much less heating. So the presence of liquid water appeared to be an all-or-nothing proposition: if there was significant liquid anywhere, there had to be a global ocean. Moreover, if an ocean existed recently, it had existed for a long time. And if it had existed for a long time until recently, there must be one now: it would be surprising if it froze solid just before our spacecraft arrived there.

Those early conservative estimates gave heating rates of only about 1,600,000,000,000 watts. That would be a lot of heat in your electric oven, but spread inside a Europa-sized body, it is not much. Consider this: We know the heating has been going on for a long time, so there is a balance between the heat produced inside and the energy radiated out from the surface. The same thing happens in your 100-watt light bulb: Inside the bulb, 100 watts of heat are generated, and 100 watts of energy radiate out. Given the size of Europa, the tidal heat spread over the entire surface would radiate out at a rate of about 0.05 watts per square meter of surface, assuming the old conservative estimates. 0.05 watts is an amount too small to measure.

We do have measurements of the heat escaping from neighboring Io, most of which must be generated by tidal friction and radiates from a few very hot spots associated with the largest volcanoes. We can use that measurement to estimate Europa's tidal heating rate by scaling Io's rate to account for Europa's different orbit (farther from Jupiter) and size (smaller than Io). On that basis we can estimate that about 0.2 watts/m^2 must be continually removed through the surface of the satellite, which is nearly 4 times the early, conservative estimate. Accounting for the likely material properties of Europa, relative to Io, raises the Europa heat flow even more, to about 0.3 watts/m^2.

All that heat must make its way out through the ocean, through the ice, and to the surface. We can calculate how thin the ice must be in order to let the heat out. First assume that heat is carried outward through the ice by thermal conduction, like the heat moving out through the side of your coffee cup from the hot liquid inside. The ice would need to be 2 km thick to transport 0.3 watts through each square meter of the surface. If the ice were thicker, heat would build up inside, and the ice would melt until it reached the "steady state" thickness of 2 km, which would then conduct the 0.3 watts.

In this conductive model, the ice thickness varies inversely with the amount of heat being transported. If the internal heating rate decreased, the ice would thicken until the rate of heat transport decreased into a new equilibrium. If the tidal dissipation were as slow as the old post-Voyager guesses (only 0.05 watts/m^2), the ice would be about 12 km thick. Yet even that value is still quite thin compared with the 150-km depth of the ocean.

The ice could be much thicker than that and still transport the same amount of heat if the ice were convecting instead of conducting, because convection carries heat more quickly and efficiently. Convection would require the ice itself to flow vertically, carrying heat up and then sinking to pick up more heat. The ability of solid ice to flow in that way depends on very exact required properties of the ice, especially its viscosity, which describes its flow characteristics. But we do not know those properties. Results depend on assumptions. Most estimates require ice thicker than 20 or 30 km for convection to occur. Some quantitative modeling even suggests that any convecting layer would probably be inconsistent with the existence of a liquid ocean, a result that strongly favors the thin conducting ice layer.

For a given rate of internal heat production, if the ice layer were transporting the heat out by efficient convection, it would be thicker than if the transport were by relatively inefficient conduction. Thus, either a thin-ice model (thinner than 10 km) with conductive heat transport or a thick-ice model (thicker than 20 or 30 km) with convective heat transport could be consistent with a rate of heat flow in the middle of the range of plausible values.

If the tidal heating were near the old low guesses, the ice would be thick enough that convection would have a good chance of being activated, letting the ice be substantially thicker than 20 km. This thickness was the canonical value at the time of Galileo's arrival at Jupiter. Consequently, the initial authorized interpretations of Europa's geology were based on the notion that the processes were dominated by solid-state convection and other solid-state flows in the ice. In that view, widely publicized and promoted under the authority of the Galileo project, the ocean would have been isolated from the observable surface and could not play a direct role in the geological processes that shaped the surface.

By itself, modeling of heat transport on Europa is too uncertain to definitively discriminate between thin conductive or thick convective ice. Europa's actual current physical condition may depend on its thermal history. If the heating rate has increased toward its current value, the ice would likely be thick and convecting. If the heating rate has decreased from significantly higher rates, then the current ice would more likely be thin and conducting. Because tidal heating is driven by orbital eccentricity, the tide-driven history of Europa's orbit may be crucial.

Contrary to the official story, what we see on Europa suggests that the ocean has been directly linked to the surface, and that result is consistent with thinner ice than allowed by convection. What first convinced me was consideration of the observed tectonic features. Remember all those cracks and ridges? To understand them, we need to consider how tides stress the icy shell.

***O *

The history of stress on the icy crust is recorded by the cracks that dominate more than half the surface. This record, as laid out in images, includes the entire geological history of the satellite. Unfortunately, that is not saying much if "geological history" is defined as the period during which the currently visible surface was created. In fact, the visible surface records only the most recent 1 percent (or some 50 million years) of the 4.6-billion-year age of Europa. Moreover, even during the brief lifespan of the surface we have imaged, the continual slicing and dicing have left an interpretable record covering only the last few million years. Nevertheless, Ockham's Razor implies that the processes recorded in the most recent tectonics represent the same types of continual tectonic modification that have gone on for much of Europa's history. There is no compelling reason to believe otherwise.

In order to consider how tidal stress would affect the surface, we need to marshal our best understanding of the likely physical properties of the ice. The upper part of the ice crust is very cold (about –170°C) because heat that reaches it from the interior is rapidly radiated into space. So this upper portion of the ice is brittle and elastic. The elasticity means that it acts springy, while the brittleness refers to its breakability when a certain stress limit (its strength) is exceeded. Near the bottom of the crust, the ice is much warmer, reaching the melting point (about 0°C) where it meets the ocean. The warmer ice is a solid, but it is viscous, capable of flowing slowly like cold honey. A more apt analogy is a glacier, a river of slowly flowing ice.

In between there is a transition from brittle-elastic properties near the cold surface to a predominantly viscous character near the warm base of the ice. The behavior of the ice also depends on how fast it is distorted. When ice is distorted (or "strained") quickly, it behaves elastically and may break in a brittle fashion, but when it is distorted slowly, it flows. A good example of a material that can display the same range of behavior is Silly Putty. Stretch it quickly and it is an elastic band, or very quickly and it acts brittle and cracks; but if you pull it slowly, or just let it settle, it flows as a viscous fluid. Also like ice, warmer Silly Putty tends toward viscous behavior, and colder Silly Putty will tend to be more elastic and breakable.

On Europa, as the ice crust rides over the changing elongated shape of the ocean, the cold brittle-elastic surface stretches and compresses and shears elastically, depending on where it is. This distortion results in a corresponding stress within the material. (In materials science, the word *strain* means the geometrical distortion, while *stress* represents the internal forces in the material. Stress and strain are two distinct, but mutually dependent, things.) Below the elastic part, a few more kilometers of warmer, viscous ice follow along, but do not build up elastic stress.

In order to compute the stress field in the surface ice, we model the crust as an elastic sheet overlying the tidally deforming shape below it. For a mental picture of this physics problem, imagine a thin rubber sheath stretched over a slippery spheroid, which changes shape and elongation underneath it. The stress in the rubber will continually change in response to the changing shape of the slippery body that it encases.

This model involves various assumptions that are reasonable on a global scale. For example, we assume that the ice sheet is continuous and uniform in its elastic properties. We must assume too that the elastic properties of the ice are similar to those measured in laboratories on Earth. In detail these approximations are surely inexact, but as a first step toward understanding global stress fields, they are reasonable. Like all physicists, we would rather work with a spherical cow than a real one. Knowing how tides affect the stress on such a crust, we can interpret the tectonic record of observed cracks and ridges.

As the Galileo images began to come to Earth in the late 1990s, we realized that the data amounted to a record of tectonic features that was detailed enough to compare with widely accepted theoretical models of stress. I knew exactly how to perform the stress calculations to test the new incoming data against these models. In fact, that is precisely why I was on the Galileo imaging team. I had made the case in my 1976 proposal that what would be seen at the Galilean satellites would largely depend on tides, and that I would know how to make the connections. And now, at last, we had the images in hand.

But my hands were tied. As described earlier, the arrangement within the Galileo imaging team was that those who did the image sequence planning would have first dibs on reporting discoveries and publishing results. This system ensured that the most politically powerful team members would control the science analysis. My research group at the University of Arizona had been locked out of the sequence planning, so we were set up to be benchwarmers in the data analysis stage as well.

The key geological power players on the imaging team, Ron Greeley of Arizona State University and Jim Head of Brown, divvied up the sequence planning, and thus the data analysis. Between them, they took control of alternating encounters with Europa, each of which covered only a portion of the surface with any detail. I found it curious that discoveries would be reported and documented and interpreted in parcels that were based on separate encounters with the satellite. It seemed to me that reconstructing the story of Europa would depend on an integrative approach. Certainly, with tides as an underlying force, we needed to be addressing the data by thinking on a global scale, which anyway was my natural tendency as a celestial mechanician. However, both Greeley and Head were geologists, and they followed the geologist's tradition of working first with particular regions or subsets of the information. That approach is not necessarily bad, but as my student Randy Tufts, an experienced geologist himself, used to tell me, the problem was not that geology is bad, it was that we were seeing it done badly.

Whether the science was good or not, the political arrangement that locked out me and my students meant that Greeley and Head each would control the press reports, the scientific conference presentations, and the initial publications in the scientific literature. Their initial impressions of the images would become the authorized results from the Galileo mission as reported to the world beyond the imaging team.

Greeley and Head assigned much of the work on image-sequence planning and scientific analysis to their students and post-doctoral associates. Eventually, one of those people, Robert Pappalardo, was anointed as a key spokesman regarding Europa for the Galileo project. As a former student of Greeley's at ASU and then a post-doctoral associate of Head's at Brown, he was doubly well connected with those in power.

My original 1976 proposal to NASA was based on the idea that tides on Europa might be important, and that my interpretation of the images might add something of value to traditional photo-geology. The notion that tidal forces played a critical role had been proven (at least for Io) when Voyager flew past the Jupiter system in 1979. By the mid-1990s, it was evident that interpretation of what we saw at Europa would also require understanding of the effects of tides. I felt, more than ever, that there would still be an opportunity for me and my students and associates to figure out what processes might lie behind the appearance of this strange satellite. So my group and I waited impatiently for the initial interpretation by Greeley, Head, and Pappalardo to be completed. We were sure we could bring a lot to the interpretive table once we were given the chance.

Although I had been appointed to the team for my expertise on tides, they chose not to invite me to collaborate, keeping me locked out of the initial science interpretation. Instead, Pappalardo asked Paul Helfenstein, who in the mid-1980s, as a student at Brown University, worked with his professor, Marc Parmentier, on calculations of tidal stress. In addition to being smart, Paul is one of the nicest people in planetary science. He agreed to give Pappalardo the software that he had developed to calculate the tidal stress due to non-synchronous rotation.

Eventually, my group did get into the act. It turned out, ironically, that the software Helfenstein gave Pappalardo was in an obsolete medium, a magnetic tape format for which no reader still existed. My students Greg Hoppa and Randy Tufts were monitoring the situation, partly because they too were irritated about being excluded, but also because I had shown them how to do the stress calculations from scratch. Finally, once Pappalardo was convinced that no usable tape reader existed, he allowed my group to proceed with the analysis. Greg got to work and within a few weeks returned to me with a complete set of calculations of tidal stress. It was that analysis that eventually became the basis for explaining most of the major tectonic patterns on Europa.

Fig. 5.1 shows a typical stress map which graphically represents Greg Hoppa's 1997 calculations. It uses a Mercator projection, the same type usually used to make flat world maps of the Earth, with locations given by latitude (up and down on the map) and longitude (right and left). As on the Earth, the location of longitude zero is decided by convention. On Earth it is the site of the Royal Observatory at Greenwich, England. On Europa, it is the center of the Jupiter-facing hemisphere. For orientation, the locations of a few of the major global lines are indicated, including Udaeus Linea and Minos Linea (c.f. **Fig. 2.4** or **Color Plate 2**), as well as Astypalaea and Agenor, which will also be major players later in our story. The short lines represent the strength and direction of the stress (tension or compression) at any point on the surface.

The stress pattern here is just an example of what these plots look like. In this case, it shows the stress that would be expected about 1/8 of an orbital period after the perijove point in the orbit, with a little extra rotation of Europa thrown in. The point is that as Europa moves its body changes shape,

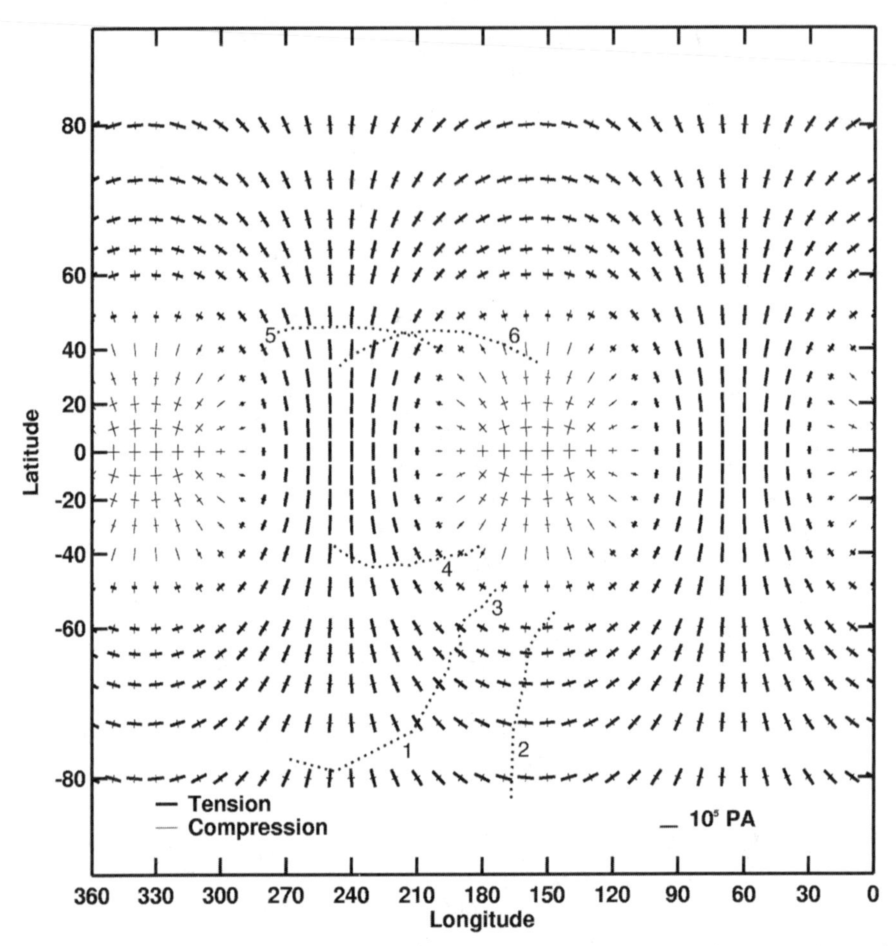

FIGURE 5.1: An example of one of our stress maps. Crossed lines indicate the orientation and magnitude of the principal stress components. Bold lines indicate compression and fine lines tension. The length of each stress line shows its magnitude (according to the scale bar at the bottom). Typical stresses are around 10^5 Pascals, which is similar to the atmosphere's pressure on Earth. The locations of several major Europan lineaments are indicated by dotted lines: (1) Astypalaea Linea; (2) Thynia Linea; (3) Libya Linea; (4) Agenor linea; (5) Udaeus Linea; (6) Minos Linea.

so the elastic skin sliding on its surface becomes stressed. In this case **Fig. 5.1** shows a lot of compression in all directions near the equator around longitudes 0° and 180°. That makes sense because the tidal elongation of the body would be decreasing as Europa moves farther from Jupiter. At the same time, 90° away, the stress would mostly be tension in a north-south direction, as also shown in **Fig. 5.1**.

This diagram is complicated, but the main thing I want you to notice is fairly simple: The major linear features observed on Europa (the dotted lines) tend to be perpendicular to the direction of tension (the dark lines), exactly what you might expect if the observed features formed when tidal tension cracked the ice. This idea turned out to be the key that allowed us to explain the distinctive crack patterns on Europa and to decipher the record of its history.

Years later, even having seen our publications, Pappalardo was still trying to figure out how to calculate tidal stress. At times, to my surprise, he even presents Greg's results as his own. He can get away with a lot. Such is the power of the mantle of authority, once bestowed by the Galileo project.

Like most planetary bodies, Europa probably formed with a spin that was not synchronous, perhaps not even close to synchronous, with its orbit around Jupiter. For each single orbital trip around the huge planet, the newly forming moon probably spun on its axis, or rotated, many more times than once. The rate of Europa's rotation early in its life reflected the result of the impacts that occurred as other smaller bodies, also in orbit around Jupiter, accreted to the growing satellite. The tidal torque would then have slowed the spin to nearly synchronous (at a 1:1 ratio) with its orbit. This process would have taken about 100,000 years, a blink of an eye compared with the 4.6-billion-year age of the solar system. The rotation would have become precisely synchronous if Europa's orbit were perfectly circular.

But it is not perfectly circular, and in the mid 1980s I pointed out that, with an eccentric orbit, tides may cause the spin to settle at a rate slightly faster than synchronous. And Europa is so warm that it probably cannot support a concentration of extra dense rock, like the one inside our own Moon's Earthward sector, that might have helped maintain synchronous rotation. So, it's the combination of a warm interior and an eccentric orbit that makes non-synchronous rotation plausible. Both are effects of the Laplace resonance. First, the resonance drives the orbital eccentricity. Second, the tidal friction results in the internal heating.

Still, we don't know for sure whether the rotation is non-synchronous, Theory alone cannot tell us, because the strength of the tidal torque is so uncertain. Direct observations, whether by telescope or spacecraft, don't help much either, because we don't have adequate resolution over a long enough time to see any change. But as we will see in later chapters, much of the tectonic record seen on Europa's surface could be accounted for by non-synchronous rotation.

Over the long term, tens of millions of years or more, tidal deformation of

Europa contributes to changes in the resonant orbits of the Galilean satellites, which in turn modifies orbital eccentricities, which changes the strength of the diurnal tides.

In the resonance, the periods of the orbits have their 1:2:4 ratio, and the satellites always align in pairs at specific points in their orbits. As Laplace showed, this configuration is stable, maintained by the tiny gravitational pull of each satellite on each other one. Ordinarily, a satellite's effects on the others would be minimal. However, because the geometry is so repetitious and periodic, the effects are cumulative and strong, hence the term resonance.

Consider what it means to have a pair of orbital periods in an exact ratio of 1:2. At the moment that the faster moon, on the inner orbit, passes the other, we say they are in "conjunction." After that point, the slower one falls behind. Conjunction occurs again only after the faster satellite completes its second lap, which happens to be exactly at the point at which the slowpoke is completing its first orbit. This second conjunction (and every conjunction thereafter) occurs at exactly the same place as the first conjunction did.

Remember, Io is in a 1:2 resonance with Europa, and Europa is in a 1:2 resonance with Ganymede. Europa, as the middle child, plays a key role in this ballet. Europa's orbit is eccentric (**Fig. 4.2**), bringing it closest to Jupiter at one end of its orbit (known as Europa's *perijove*) and farthest from Jupiter at the other end (its *apojove*). In the Laplace resonance, the conjunction of Europa with Io always occurs in line with Europa's apojove, and conjunction of Europa with Ganymede always occurs in line with Europa's perijove. In other words, the two conjunctions always occur at opposite ends of Europa's elongated orbit (**Fig. 5.2**).

The mutual gravitational effects among the satellites maintain this repeating geometry by adjusting the value of Europa's orbital eccentricity. The closer the periods are to the exact resonance ratio, the bigger Europa's eccentricity must be in order for the repeating geometry of the conjunctions to be maintained.

This behavior is so weird and seemingly magical that when I first heard about it in 1968 I knew that celestial mechanics was what I wanted to do. In general, various types of orbital resonances play roles in many of the most important interactions among planetary bodies, because of the way they can enhance mutual gravitational effects. Here I consider only the effects of the Laplace resonance that are most important to Europa's evolution, and I avoid the math that we usually use to describe these amazing interactions.

One effect is that Europa's eccentric orbit, forced by the resonance, drives the diurnal tides. Also, tides on Jupiter and the other satellites, as well as on

FIGURE 5.2: Conjunction of Io with Europa always occurs on exactly the opposite side of Jupiter from conjunction of Europa with Ganymede. The three satellites can never all be lined up together on the same side of Jupiter, or even come close.

Europa, can actually change the orbital motion of the satellites with major implications for the history of the resonance.

Chuck Yoder, a student of Stan Peale's at the University of California in Santa Barbara, developed a mathematical theory in 1979 that showed how the system could have evolved from a non-resonant state into the current Laplace resonance. In that scenario, Europa's orbital eccentricity would have gradually increased up to its current value. The implication was that tidal effects on Europa would have been much weaker in the past than they are now.

Yoder's scenario was widely accepted. First, from a mathematical point of view, it was an incredibly elegant model, and thus appealed to the handful of celestial mechanicians (including me) who understood it. But a seemingly fundamental problem always bothered me. Yoder's analysis depended on a balance between the effects of tides raised on Jupiter by Io, and tides raised on Io by Jupiter. (He also assumed that Europa tides were negligible in comparison, which may or may not be true.) That balance translates into a particular ratio between the amount of heat generated by tidal friction within Io and within Jupiter. Now, we know from observing Io's volcanoes just how rapidly heat is being produced there, so the model requires a balancing tidal heating rate within Jupiter. The problem is that no one has been able to figure out how there could be that much friction inside Jupiter.

For that reason, in 1982 I argued that the history of the system may not have involved the Yoder equilibrium at all. More likely, the tides on Io dominate over the tides on Jupiter. In that case, the system must be evolving away from the resonant condition, in which case Europa's eccentricity would be gradually decreasing from previously larger values, exactly the opposite of the evolution in Yoder's model.

Given that likelihood, I suggested several scenarios which would have led to the system's currently moving out of resonance. One idea was simply a primordial origin, in which the satellites entered resonance during the satellite-formation process. Stan Peale and others did not like the idea; it was too hard to abandon Yoder's theoretical model, especially because there was no compelling reason to believe that the process of satellite formation would establish a resonance. A convincing mechanism had not yet been demonstrated.

Recently, however, new ideas about the formation of the Galilean satellites have included just such a process. The revised models of formation were motivated by a problem with earlier models of satellite formation. The key issue involves the structure of Callisto, whose density profile (inferred from its effects on the Galileo spacecraft's trajectory) shows that it is only partially differentiated. It was not heated sufficiently to allow the iron, rock, and water to separate by density into distinct layers like those within Europa.

How could Callisto be like that? Even if the circum-Jovian nebula had barely enough material to build the Galilean satellites, accretion would have been so fast and the temperatures would have been so high that Callisto should have become fully differentiated. Robin Canup and Bill Ward, of the Southwest Research Institute in Boulder, have addressed this problem by proposing that the satellites formed later, near the end of Jupiter's formation period, while dust and gas were flowing from the solar nebula into Jupiter through an "accretion disk" around the giant planet. They calculated that the satellites could form slowly within the accretion disk, consistent with the amount of internal differentiation now found in each of the satellites.

Most important for the history of the orbital resonance, the gravitational interaction of the growing satellites with the accretion disk would cause their orbits to migrate. They would induce spiral waves in the accretion disk (similar to the spiral waves in galaxies or in Saturn's rings) and their gravitational interaction with the waves would change the satellites' orbits.

This migration of orbits provided a way for the Laplace resonance to get started as part of the same events that formed the satellites, as demonstrated in 2002 by Stan Peale and his post-doctoral associate Man Hoi Lee. Subsequent to the formation in resonance, tides on the satellites would modify the orbits, decreasing the orbital eccentricities from higher early values, as I had described in my 1982 paper. It was personally gratifying that my conjecture was proving plausible. Even better was that Peale and Lee acknowledged that I had it right all along.

The several scenarios that I proposed in the early 1980s all differed from Yoder's equilibrium model in that Europa's orbit would be becoming more circular, not more eccentric. Given the tidal friction apparent in Io from all that volcanism, significant change in orbits might have taken place quite recently, within the past 100 million years or so.

That time scale is strikingly similar to the probable 50-million-year age of Europa's surface. At the time of formation of the oldest features currently visible on Europa, the satellite's orbital eccentricity would have been substantially larger. All the tidal effects (heat, stress, etc.) that are so important on Europa now would likely have been, in this scenario, even stronger in the past. A natural question is whether there is evidence for change in the geological record over the past 50 million years as the eccentricity decreased.

***O *

In fact, the canonical Galileo reports on Europa's geology claimed to have found such evidence. Purported changes over time, especially in tectonic features, were presented as indications of a thickening of the icy crust, supposedly consistent with decreasing tidal heating. Moreover, the party line declared chaotic terrain as the youngest type of surface on Europa, formed only late in the 50-million-year age of the surface. The canonical storyline was that chaotic terrain formed from convection in the solid ice, and that the ice only recently became thick enough for convection and subsequent chaos formation. So the emerging geological party line implied a gradual decrease in tidal heating over the 50-million-year age of the current surface. That story fit perfectly with my old theory that Europa's eccentricity had been decreasing over about that same amount of time.

I should have loved that story. It was great in so many ways. For one thing it demonstrated the interrelatedness of evidence from different parts of planetary science, things as diverse as the age of Europa's surface based on comparing the small number of craters with the impact rate from comets and other small bodies, and orbital evolution theory based on the tidal heat radiating out from Io. The geology seemed to be confirming the predictions of theoretical celestial mechanics. As the leading advocate for such orbital evolution, and having taken considerable flak for it until Peale and Lee came around, I should have found it very gratifying. And the geologists were quite willing to cite and credit my predictions. If I had kept my mouth shut, I could have been a hero.

The problem was I found the geological arguments unconvincing. The geological time sequences seemed to be based on ad hoc interpretations of the images. Geological mapping exercises that purported to show changes in tectonic style or thermal processing were ambiguous and subjective.

The argument based on chaotic terrain made the least sense to me. As my students and I studied the images, we saw no evidence that chaotic terrain was especially recent. To be sure, Conamara Chaos (**Fig. 2.7**) was a very fresh feature, but it was selected as a target for detailed imaging precisely because it was so fresh and obvious. Yet even there, we could see the cracks and ridges that were already beginning to cross it (**Fig. 2.9**). As we surveyed the surface, we could see that over much of the surface, hidden under densely packed strands of newer ridges, lay very old areas of chaos. The putative freshness of most chaotic terrain was an observational artifact: It is simply easiest to spot the examples of chaotic terrain that formed most recently.

Moreover, even if chaotic terrain formation were only a recent process, the argument that it implied thickening of the ice over time seemed completely backwards. The appearance of chaotic terrain, with rafts that evidently drifted around to new positions, seemed to argue for melt-through exposure of the ocean. That interpretation would suggest that the ice had gotten thinner with time, not thicker.

In fact, even before I had gotten into the fray, there was a battle among the imaging team geologists over this point. The amazing pictures of Conamara Chaos were a major early accomplishment. The veteran planetary geologist Mike Carr of the US Geological Survey was the team member assigned to lead the preparation of a paper for the influential and prestigious journal *Nature* reporting on this discovery. Along with most people who looked at those images, Carr interpreted Conamara as a site of relatively recent melt-through of the ice, exposing the ocean below. For a brief period in early 1977, that was the official story for the media. For example, in April 1977, Carr and project scientist Torrence Johnson described Conamara in those terms during on interview on the PBS MacNeil-Lehrer news show. And that is how Conamara was described in Carr's early drafts of the *Nature* paper.

In the same time frame, however, Bob Pappalardo and his mentor Jim Head had begun to promote their view that everything on Europa involved thick ice and solid-state processes. The keystone of their case was the putative existence of a class of features they called "pits, spots, and domes," although that taxonomy has since proven to be premature and its generalizations incorrect (see **Chapter 14**). With the backing of his mentors, Pappalardo was assigned to lead the preparation of an imaging-team paper arguing for solid-state convection in thick ice. The plan was for Pappalardo's paper to appear as part of a set of papers in *Nature* that would present the early discoveries of the Galileo mission. The problem was that his supposed evidence for thick ice was in direct contradiction to Carr's interpretation of

Conamara, which was to be published in the same issue of *Nature*. Having become the spokesperson for the imaging team regarding Europa, Pappalardo had a platform of authority from which to promote his own views. Considerable pressure was brought to bear during the preparation of the Carr *et al.* paper, so that, while portions of it reflect the interpretation of oceanic exposure, the paper finally favors a vaguely described solid-state formation process for chaotic terrain. From that time forward, thick ice became the obligatory canonical model.

Pappalardo expressed surprise that I did not jump on the thick-ice bandwagon. It would have been so good for my career. The new canonical story—that chaotic terrain was a sign of recent thickening of the ice— fit perfectly with my early calculation that tidal heating must have decreased with time. My role as the team's celestial mechanics guru would have been established. Unfortunately, I just could not buy the supposed geological evidence that the ice was so thick that the ocean was cut-off from the surface. Not only did I reject my appointed role in the official Galileo narrative, but I was stepping on the toes of the geologists. My life would have been so much easier if I had gone with the canonical flow, but being me I had to follow the evidence where it took me.

The fact that there is no clear evidence for change in geological processes over the recorded geological history does not mean the orbits did not evolve. Evidence for such change may have been destroyed by more recent resurfacing, or perhaps it is too subtle to recognize. Equally plausible is that the geological processes recorded on the surface have been continuous over the surface history, and continue currently. In lieu of other evidence, Ockham's Razor favors that simple interpretation.

Studies of orbital dynamics are continuing. We need to understand how tides raised on Europa and Ganymede may have affected the evolution, especially if we take into account the accompanying changes in the geophysical states of the satellites, and how they feed back into the orbital evolution. Historical measurements of orbital periods might help. We have a useful record of observations that goes back to Galileo's original seventeenth-century data on timing the orbits, and Kaare Aksnes (University of Oslo) and Fred Franklin (Harvard-Smithsonian Center for Astrophysics) have been investigating the evidence for changing orbits.

These orbital studies, along with detailed investigation of what is really seen on Europa, may eventually play key roles in clarifying the geological and geophysical history. We may also be able to make predictions for the future. Europa has had an ocean and an active surface in the recent past, probably continuing up to the present. In the future, it may well freeze solid, depending on how its orbit continues to evolve.

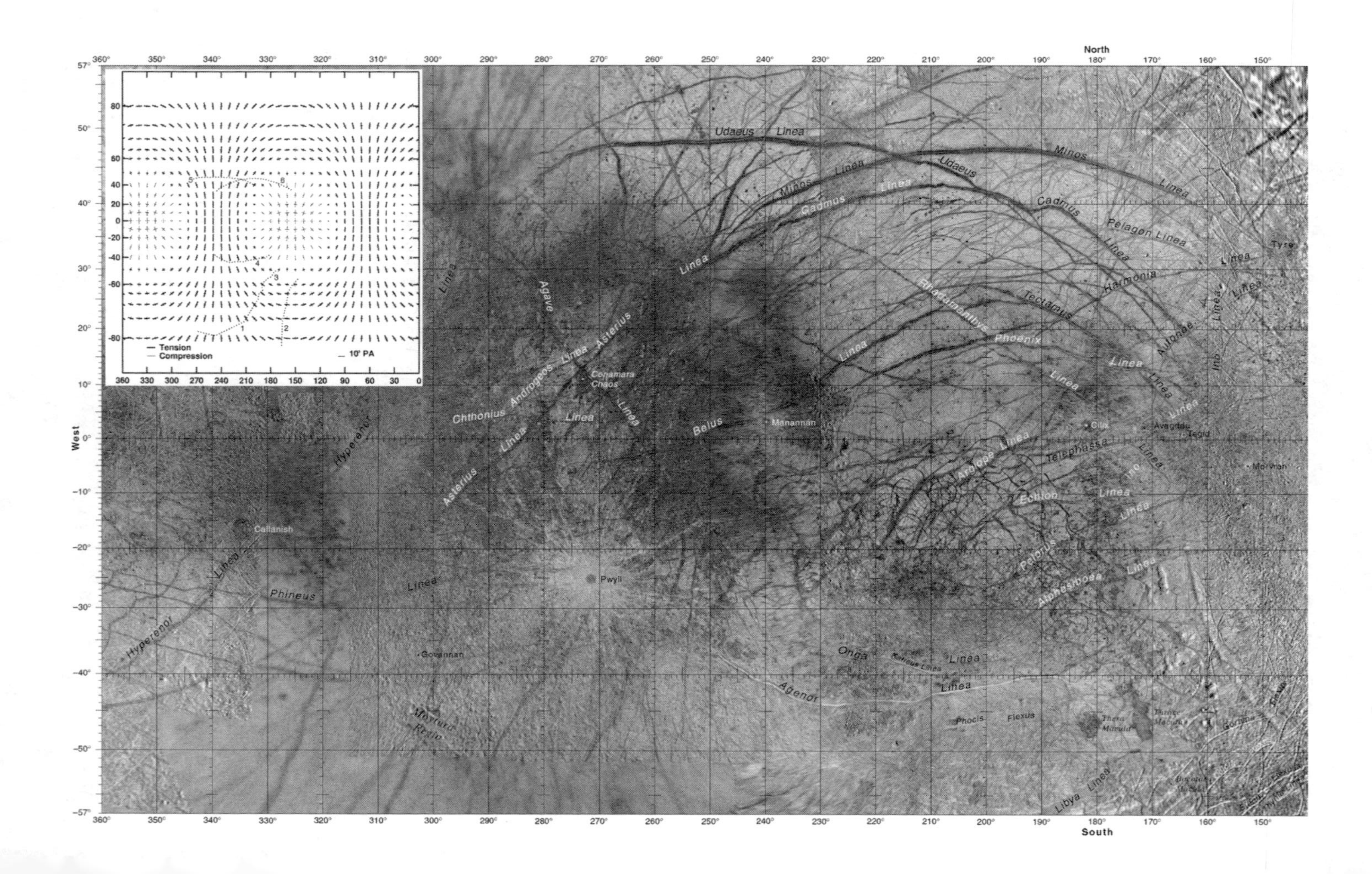
North
South
West
Udaeus Linea
Minos Linea
Cadmus Linea
Pelagon Linea
Harmonia
Tectamus
Rhadamanthys
Phoenix
Autonoe
Ino
Tyre
Agave
Asterius
Androgeos
Chthonius
Conamara Chaos
Belus
Manannán
Cilix
Avagddu
Tegid
Morvran
Telephassa
Argiope
Echion
Pelorus
Alphesiboea
Onga
Agenor
Phocis
Flexus
Thera Macula
Thrace Macula
Libya Linea
Pwyll
Govannan
Callanish
Phineus
Hyperenor
Moytura Regio
Tension
Compression
10° PA

Global Crack Patterns

6

The first direct connection between Europa geology and tides was made by Paul Helfenstein and Marc Parmentier at Brown University in the early 1980s. They demonstrated that the large-scale linear features on Europa are roughly perpendicular to the direction of the maximum tidal tension in the crust, a relationship touched upon in the last chapter, and evident in the beautiful global mosaic map produced after the Galileo mission by the US Geological Survey (**Fig. 6.1**).

You would expect a crack to form perpendicular to the tidal tension, so this general correlation of lines with stress tells us that these lines probably represent cracks. Yet most of these linear features do not look like cracks. In our tour of the surface we saw that they are usually "triple-bands," which on close inspection prove to be composed of ridges with dark margins. Moreover, the common denominator of all the ridges is the ridge pair. So it seems that each double ridge marks a hidden crack, probably in the valley between the ridges. The correlation is an important clue to how double ridges and adjacent surface darkening (the "triple bands") develop.

Looking more carefully at the crack patterns, we can use them as a record of Europa's past. In 1986, Alfred McEwen mapped many of Europa's surface lines using Voyager images. McEwen was then a graduate student at Arizona State University and also working at the U.S. Geological Survey's Astrogeology center in Flagstaff, Arizona. He became a member of the

FIGURE 6.1 (opposite page): A map in a Mercator projection of Europa, constructed by the USGS from a mosaic of mostly Galileo (and some Voyager) images. The major lineament systems are oriented roughly perpendicular to the tension predicted from tidal stress calculations (**Fig. 5.1**, reproduced here and inset at the upper left for comparison). In the USGS map, the two dark arcs near about 45°N are Udaeus and Minos Linea, which are labeled 5 and 6 in the inset from **Fig. 5.1**. The names of lineaments are drawn from Homer's Odyssey, related to Europa in classical mythology, and other features are named after Celtic myths, heroes and place names, all as assigned by the International Astronomical Union (IAU) Nomenclature Committee.

Galileo Imaging Team, and is currently my colleague on the Planetary Sciences faculty at the University of Arizona, where he directs the Planetary Image Research Laboratory, whose students and staff have produced some of the most spectacular and revealing versions of Galileo images.

What McEwen found was evidence that supported my idea that Europa might rotate nonsynchronously. He noted that many cracks are about 25° too far east relative to the expected stress patterns. For example, if we moved Minos Linea 25° further west (**Fig. 6.1**) it would fit the tension pattern much better. Evidently, after tidal stress caused the cracking, further rotation of Europa carried the cracks eastward (relative to the direction of Jupiter) to their current positions. I could quibble with some details of McEwen's paper, but the main principle is crucial: The crack record may potentially be deciphered to determine the relative age of individual features, based on how much further west they must have formed. More generally, the crack patterns might be exploited to investigate the history of the satellite. And the patterns from the Voyager images seemed to support my idea that Europa might be spinning nonsynchronously.

So as the early Galileo images began to arrive a decade later, we were primed to look for more evidence of Europa's rotational history. The first systematic evidence for a time sequence of cracking that could be correlated with nonsynchronous rotation resulted from Paul Geissler's expertise at multi-spectral remote sensing, combined with his ability to link interesting observations with fundamental theory. To consider the images that Paul used, it may be helpful to understand how the Galileo spacecraft got them.

During each of its orbits around Jupiter, the spacecraft generally spent a good fraction of its time very far out, and then it swooped into the neighborhood of the satellites' orbits. Generally the satellites' positions were far apart, so the Galileo craft did not fly close to more than one satellite during each orbit. The swoops came to be labeled not only with the number of the orbit, starting with the first one after the spacecraft was inserted into orbit around Jupiter, but also with a letter identifying which satellite was encountered on that orbit. Thus the first orbit was called G1 because it involved a close targeted encounter with Ganymede. E2 was the first encounter with Europa. The letter is redundant information, since only one satellite was targeted on each of Galileo's orbits, but the E or G serves as a reminder of which satellite was favored on each swoop.

With orbit G1 targeted on Ganymede, Galileo did not get close enough to Europa to get high-resolution images, but it did get a set of images at a resolution of about 1.6 km per pixel covering a wide region around the Udaeus-Minos intersection. The same region was imaged several times using the same black-and-white camera, but each time through a different wavelength filter. Each image recorded the brightness in a certain range of wavelengths of light, many beyond the wavelengths visible to the human eye. That is, many were at wavelengths even longer than red, in the infra-red.

These images provided just the kinds of data Paul Geissler loves. He obsessively combined the data collected through different filters to produce a stunning variety of color composites. Usually a night worker, Paul was so excited that he was up all day perfecting these beautiful images. But while they were pleasingly colorful, they did not show what the human eye would have seen on Europa's surface, which in reality would appear white and bland. Instead, in order to get at the information that interested him in the image data, Paul had exaggerated the color and contrast and as needed made color substitutions. So, for example, some color pictures displayed as a combination of red, green, and blue (the standard components of color TV) represented combined images that in actuality were taken through filters at infrared, red and green wavelengths, respectively. **Color Plate 2** is one of these products.

As a result, Paul's images, in different versions combining filters in a variety of ways, showed Europa's ice in various shades of pink and blue, with the dark lines bordering the triple bands in shades of orangey brown. When I had first hired Paul to help with one of Galileo's quick encounters with Earth on its way to Jupiter, he did the same thing with images of Antarctica, taken from the spacecraft with the very camera later used to image Europa. For Antarctica, he had discovered that in his false-color images the pinks and blues could distinguish between the sizes of the grains of ice in snow and clouds. For Europa, Paul could tell us how the ice-grain sizes varied from place-to-place.

And, for us, there was an even greater payoff: His enhanced images showed something that got my whole research group excited. Paul noticed that he could easily distinguish which lineaments appeared to cross over on top of other ones. These show up clearly, for example, in **Color Plate 2**. So, these cross-cutting relations revealed the order in which the linear features were created. Invoking Ockham's razor we made the simple assumption: Older lines mean older cracks.

Then Paul noticed some patterns. The older the dark colored markings, the more they appeared faded. With increasing age, the dark margins of the "triple bands" seemed to revert to the coloration of the surrounding terrain (e.g., **Color Plate 2**). Paul found that the oldest lineaments are fairly bright and white. The intermediate-age ones are the triple bands with their dark margins. And the youngest were faint thin lines, which we interpreted as cracks that had not yet had time to develop ridges or dark margins.

The most striking pattern was an orderly sequence of orientations that changed systematically with age. The most recent lineaments trend roughly northwest to southeast, older ones run more nearly east-west, and the oldest run southwest to northeast. Going forward in time, the orientations had systematically rotated clockwise. Moreover, the trend was followed even among the intermediate-aged "triple-bands." Examining the intersection of the triple-bands Udaeus and Minos near the center of the region, we can see that Udaeus crosses over Minos, part of the general trend: The younger feature is oriented further clockwise than the older one. The changing orientation of lineaments inferred from Galileo's first orbit (G1) suggested a systematic clockwise rotation of the direction of tension in the Udaeus-Minos region over time.

At one of our weekly research group meetings, where my students and post-doctoral associates discussed their progress and plans, Paul insisted that we needed to compare this clockwise trend with the tidal stress plots that Greg Hoppa had been computing. It was a very good idea, so we sketched the positions of Udaeus and Minos on a stress plot (similar to **Fig. 5.1**, inset in **Fig. 6.1**). A remarkable pattern jumped out from this comparison: If the real estate in the G1 images had moved from the west to the east over the last few tens of degrees in longitude to its current location, it crossed a part of the stress diagram where the direction of the local tidal tension would have rotated in the clockwise sense, changing its orientation just as the observed record of lineament orientations had changed. (The tension would have changed over a range of directions perpendicular to the directions of the lineaments Paul had mapped, but to more or less the same extent.)

This discovery provided additional compelling evidence for nonsynchronous rotation. Apparently, in the past, the Udaeus-Minos region had been further west, where tidal stress caused the older cracks to form. Later, as Europa rotated nonsynchronously, this region, along with the entire surface, moved eastward. That is, it moved eastward relative to the direction of Jupiter (and by definition it moved eastward relative to the longitude system). And because the stress field was generated by Jupiter, as the real estate around Udaeus and Minos moved eastward, it experienced ever-changing tidal stress. So the more recent cracks have a different orientation than the older ones.

This discovery made for a very exciting debate in my lab. Paul and my students Greg Hoppa and Randy Tufts overcame every objection I could raise and convinced me that we had evidence for nonsynchronous rotation. The result was quickly accepted within the Galileo imaging team, and then more widely after we described it in a couple of papers for which Paul was the lead author. It was so compelling because the data showed several sets of lineaments that were progressively and nearly continuously more clockwise with decreasing age, over a few tens of degrees of rotation. The appeal of the interpretation lay in its elegant linkage of several lines of quantitative research. It confirmed my own much earlier predictions, based on tidal theory, that Europa might be rotating nonsynchronously; it clearly corresponded with the beautiful images Paul produced using state-of-the-art image processing and quantitative multi-spectral analysis; and it resulted in the discovery of unexpected systematic trends that fit Greg's computed tidal stress field.

I had several reservations about this model, though, and one of them later turned out to be important. It seemed to me that once a crack had formed in the Udaeus-Minos region (or any place else), it would have relieved the stress. You would not expect more cracking in the neighborhood until either the first crack froze shut or the direction of the tension changed a great deal. Both of those things would take a long time. Once a crack forms, it must be worked continually by diurnal tides. And as I describe in the next chapter, cracks must remain open and active long enough to build the ubiquitous ridges. It would take a long time before a crack would freeze shut and allow tension to build up again. So, even as I supported Paul's story, I worried about how so many cracks could have formed with orientations that changed only gradually from one to the next.

The lines that were visible in those early low-resolution images from G1 must be among the youngest on the surface. Paul's study of changes in color over time showed that older features blend back into the background. Given that Europa's surface is less than about 50 million years old, and that the lines involved in Paul's rotation analysis are much younger, Europa's nonsynchronous rotation period (the period relative to the direction of Jupiter) would seem to be less than a few million years. A Jupiter-facing hemisphere would shift away to a different direction in less than a million years or so. That constraint on our estimate of rotation speed is not very tight; It could be much faster, but not much slower.

What other constraints are there on how we would estimate the speed of rotation? It must be slow enough for ridges to grow along each crack. Otherwise, if the rotation were too fast, the crack would quickly move east to a place where the stress is different from where it formed, and so it would freeze shut before it could grow ridges. Based on estimates of how fast ridges grow (in the next chapter), the rotation period cannot be less than about 100,000 years.

So far we had constrained the rotation period to less than a few million years and more than some 100,000 years—long enough for ridges to grow.

Could we be more precise than that? Something that I had in mind from the beginning of the Galileo project was to try to measure the satellites' rotation by direct observation. In principle, if we take a picture of a satellite at two different times, we should be able to see if it rotates. In practice there are problems. In a photo taken from a moving spacecraft, the exact position of a reference feature on the surface may be difficult to define.

Shortly after Greg Hoppa asked to work with me as a PhD student, I suggested that we try to solve this problem. We decided that the best strategy was to compare Voyager and Galileo images, because they would give us the greatest interval between observations, and so the best chance to see motion. We also decided to use the terminator (the visible boundary between the dark night side and the bright day side) of Europa as a position reference. Its direction in space was well-defined by the precisely known position of Europa relative to the Sun. With Paul Geissler's help, Greg found moderate-resolution images of a particular region taken by both spacecraft 18 years apart, each of which showed the position of the terminator at the time the picture was taken (**Fig. 6.2**).

We knew the terminator position at the time of the Voyager image, and we knew the exact time interval between the two images. So we could predict where the terminator should have been on the Galileo image if it had been rotating synchronously. The analysis presented some technical challenges, but by the time Greg was done, he concluded that the terminator was exactly as predicted. The evidence was consistent with synchronous rotation. Greg and I were dejected; We had really been hoping that we would detect something more interesting.

Paul Geissler pointed out that all was not lost if we took into account the uncertainty in Greg's measurements. No measurement is exact. In fact the terminator is fairly wiggly due to the effect of topography: ridges and bumps cast shadows that distort the day-night boundary. Greg calculated the uncertainty in the measured rotation to be about 0.5°. In other words, Europa might have rotated as much as an extra half degree in 17 years, relative to synchronous rotation, equivalent to one full rotation in about 12,000 years.

We now had the period of rotation within the range 12,000 to a few million years, while the existence of ridges probably required a period longer than about 100,000 years. Not very precise, not very certain—but that was how things stood at the end of the twentieth century, with Galileo still in orbit around Jupiter.

***○ *

The orientations of the cracks in the early G1 images had seemed like the perfect evidence to compare with theoretical calculations of stress, and thus

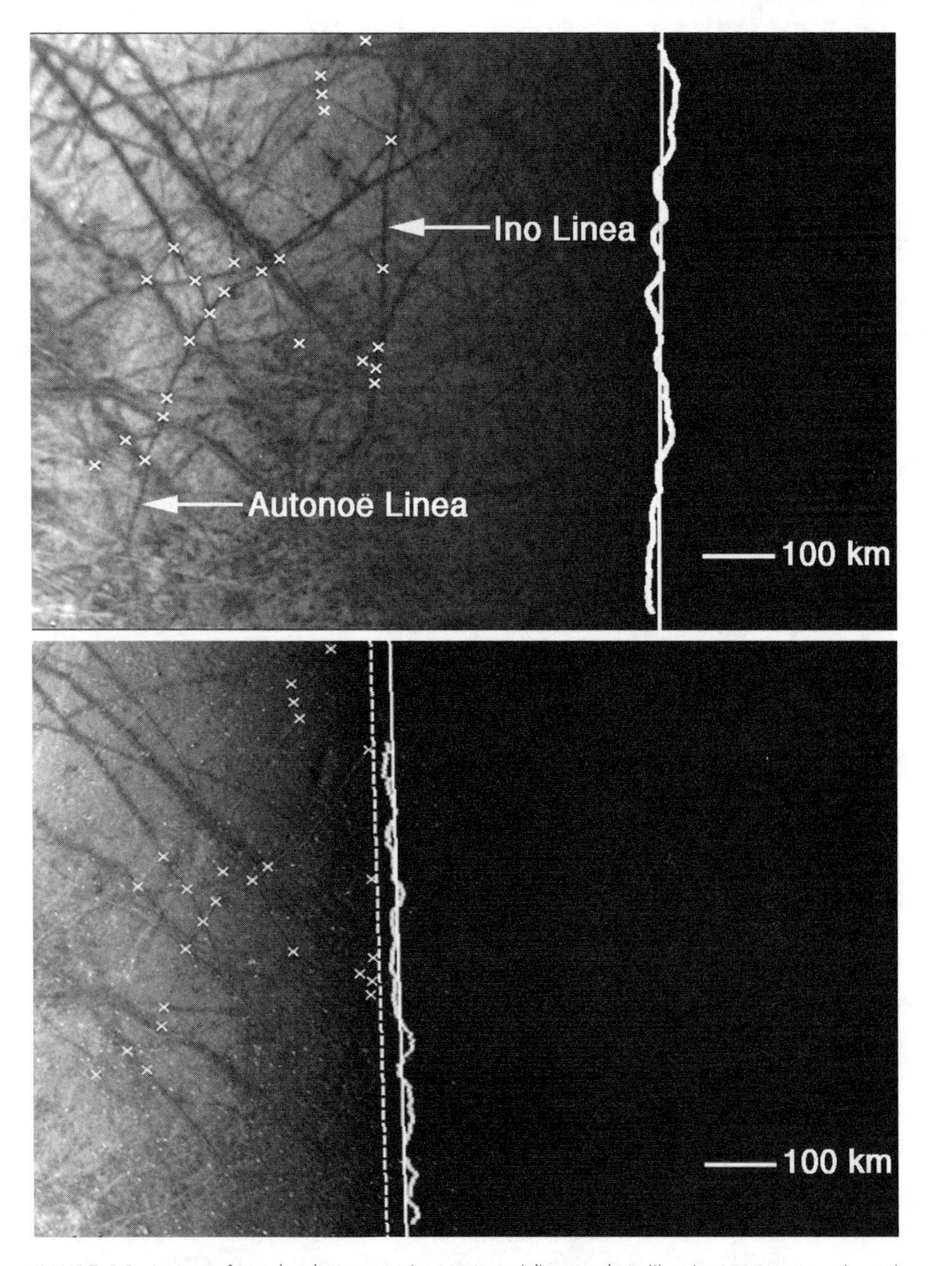

FIGURE 6.2: Images from (top) Voyager in 1979 and (bottom) Galileo in 1996 were selected by Hoppa *et al.* because they showed the same region with the terminator (day = night line) nearby. The images are reprojected to show the same viewing geometry. The position of the terminator in both images is consistent with synchronous rotation, so the nonsynchronous rotation must be very slow. This region lies between 0° and 30°N with the west edge near 172°W longitude (with features recognizable in **Fig. 6.1**).

FIGURE 6.3: Cycloidal ridges near the far south of Europa that appear prominently in a Voyager image. For location, compare this image with **Fig. 6.1**.

to infer rotation. And the systematic change in those orientations in the Udaeus-Minos discovered by Paul Geissler made for a compelling story. But a couple of years later we found an even better way to pinpoint where particular cracks had formed.

A peculiar pattern of ridges had been noted on Voyager images of Europa (**Fig. 6.3**). As viewed from above, they form chains of arcs called "cycloids," so designated because of their similarity to the geometric shape of that name. The official nomenclature of the International Astronomical Union calls them *flexi*, but it's the same thing. One of these cycloids was never named by the IAU. When my student Mike Nolan did a survey in the early 1990s of cycloid properties from the Voyager images, he labeled this one "G" in honor of his thesis advisor. Later, when Hoppa *et al.* published the paper showing how these cycloids constrained Europa's rotation period, that notation was retained, and to my students' amusement I became the only person to have had a feature on Europa named after himself.

It turns out that these distinctive crack patterns can be directly attributed to the changing stress of diurnal tides. (That discovery is described in **Chapter 12**.) In fact, with that knowledge, we can pretty much pin down where any cycloid must have formed, relative to the direction of Jupiter. In 2000, Greg Hoppa showed how this information could be used to constrain the rotation period. He considered the cycloids discovered by Voyager (**Fig. 6.3**), which appear to be relatively new, crossing over most of the other features in the area. Based on what Paul had discovered about systematic changes in the Udaeus-Minos region, we expected the cycloids to have formed in a similar fashion—an orderly west-to-east sequence over a few tens of degrees of rotation.

Greg's results *did* support the idea of nonsynchronous rotation. The shapes of cycloids required, to fit the theory, that each one must have formed at a specific place farther to the west. Those places ranged over several tens of degrees of longitude. Apparently, although they all formed in the same area on the ice, each cycloid formed at a different time while the ice moved eastward to its current position.

If that were true, the order in which these cycloids formed, as inferred from the way the ridges crossed over one another, should match the west-to-east order of their formation longitudes. In other words, a younger cycloid should have formed closer to its current longitude than an older one and, if they cross, its ridges should cross over the older one. Greg found instead that the orders were scrambled. The age sequence inferred from the ridge crossings did *not* match the order inferred from the location where they formed. This conundrum was resolved by noting that a cycloid that moved eastward to its present location may have moved through extra nonsynchronous rotation cycles. For example, a feature that formed a few degrees west may have rotated an extra 360° before it reached its current longitude. (Tidal stress is symmetrical, so that the inferred location of formation of any cycloid could be another 180° farther west, which also helps us sort out the age sequences.) So Greg was able, by invoking extra rotation periods, to reconcile the age sequence based on ridge crossings with the ages based on where they formed.

That result implied that no more than a couple of cracks form in a given region during a nonsynchronous rotation period. That rate is reasonable if each crack relieves the tidal stress in its region; Tidal stress cannot begin to accumulate again within that region until the crack anneals. That result was reasonable, but its implication was disturbing: It was inconsistent with the inferred history of the Udaeus-Minos region, where many cracks seemed to have formed in only a small fraction of a rotation.

Sweeping that problem under the rug so we could ignore it for a few years, we considered the implications of having only a couple of cracks forming in a given region in each rotation period. The region around Sidon and Delphi Flexi is an area roughly 800 × 1000 km, as shown in **Fig. 6.3.** Presumably it has been completely resurfaced in the past 50 million years, just like the rest of Europa. If a typical crack, like Sidon or Delphi yields a ridge pair about 2 km wide and 1000 km long, a minimum of ~400 cracks would be needed to resurface the area. That would take at least 200 rotations if only a couple of cracks form on each turn. To have 200 rotations in 50 million years means the period must be less than about 250,000 years.

This result allows tens of thousands of years for each new crack to remain actively worked by diurnal tides, before it moves away, freezes shut, and lets stress build up again to create the next crack. That amount of time is more than adequate for the ridges to grow by a process described in **Chapter 7**. Most significantly, this line of evidence reduces the upper limit of the nonsynchronous rotation period from millions of years to only 250,000 years. With the lower limit of 12,000 years from comparing Voyager and Galileo terminators, we conclude that the rotation period is approximately 50,000 years, give or take a factor of 5.

That analysis was based on only one region on Europa. Given how widely distributed cycloids are on Europa, and how many we have imaged, the work on the cycloids in the Astypalaea region should probably be considered only a pilot study. It demonstrates, as a proof of concept, the potential for future studies of the tectonic record to help unravel the history of Europa. When future spacecraft obtain more complete global coverage at adequate resolution and with appropriate illumination, Europa's dynamical history will be laid out and ready to read. In the meantime, the data already in hand from the Galileo mission have not yet been fully exploited in this way, and probably still have a great deal to reveal while we wait to return to the Jupiter system.

***O *

The problem remained that if only a few cracks could form during each rotation of Europa relative to the direction of Jupiter, which seemed reasonable, how could we explain our earlier contradictory results from the G1 images of the Udaeus-Minos region? That earlier work had been very convincing and was universally accepted for good reason. The model elegantly explained the several sets of lineaments that were progressively and nearly continuously further clockwise with decreasing age.

Indeed, the original approach had been so appealing that various other researchers had begun to fit the geological record into a presumed sequence of gradual, continuous changes in the orientations (sometimes called *azimuths*) of cracks. Geological mapping of various regions was done using orientations of linear features as a way to put their ages in order. Geological timelines patterned on the formats of terrestrial stratigraphy incorporated this information. (Geologists tend to use the word stratigraphy to mean a time sequence of events, because on Earth widespread layers, or strata, have built up over time, and their study has been a key to understanding geological history. There is no evidence that stratigraphy is a meaningful concept on Europa.) Various geological studies took observed linear features and structured them into systematic scenarios modeled on Paul's study of the Udaeus-Minos region. A lot of people, including some of the powerful players on the Galileo imaging team, became deeply invested in the idea of the generalized phenomenon we'd proposed—a smooth, systematic variation of crack orientations.

The problem was, we received support at a point when we no longer were sure we wanted it: Models in which numerous cracks formed during a fraction of the rotation period were precisely contrary to what we had inferred from the cycloids in the Voyager images. This inconsistency with Paul's early model was not widely appreciated, but Greg Hoppa kept reminding me about it—pressing the case for a kind of inconvenient truth. Because my own group was responsible for the evidence on both sides (the azimuthal variation in the Udaeus-Minos region and the cycloids in the Voyager images), it seemed especially important for us to resolve the contradiction. So we decided to return to the Udaeus-Minos region, where the low-resolution images from orbit G1 (**Color Plate 2**) had provided the compelling and widely accepted evidence for nonsynchronous rotation.

The earlier work had been solid research for its time, but it was based on the limited data from the first orbit, G1, when Galileo had not even come very close to Europa. Strangely, even though this same region had been imaged at much higher resolution (200 m/pixel) during orbit E15 back in the twentieth century, as late as the summer of 2002 no one had used these superior images to check what we had done with the G1 data. Compare **Fig. 6.4** with **Color Plate 2** to see how much more detail is shown by the E15 images.

Late that summer, Alyssa Sarid began her senior year at the University of Arizona. She had already done a major study of the E15 images during her junior year. (That project, and Alyssa herself, are described in **Chapter 9**.) Because she was already an expert on tectonic features in the E15 images, I asked her to help me use them to check our results from the cruder G1 data. The first step was a survey of all intersections of ridges or other linear tectonic features in the same Udaeus-Minos region that we had studied in the G1 images.

In an ideal data set, each and every crack in the selected area would cross both the next oldest and next youngest, and the appearance of the intersection would unambiguously display their order of formation. Assuming that the expected rotation of the tidal stress field (clockwise in the northern hemisphere) controlled cracking, the orientation (the *azimuth*) of the newer crack should be further clockwise than the older one. Of course there is an ambiguity. By analogy, if a weathervane changed position from north to north-east, you cannot tell if the wind direction changed by 45° or by going all the way around the other way. Similarly, if a newer lineament on Europa *seems* to be oriented further counter-clockwise, it may *actually* have formed after the stress rotated much further clockwise.

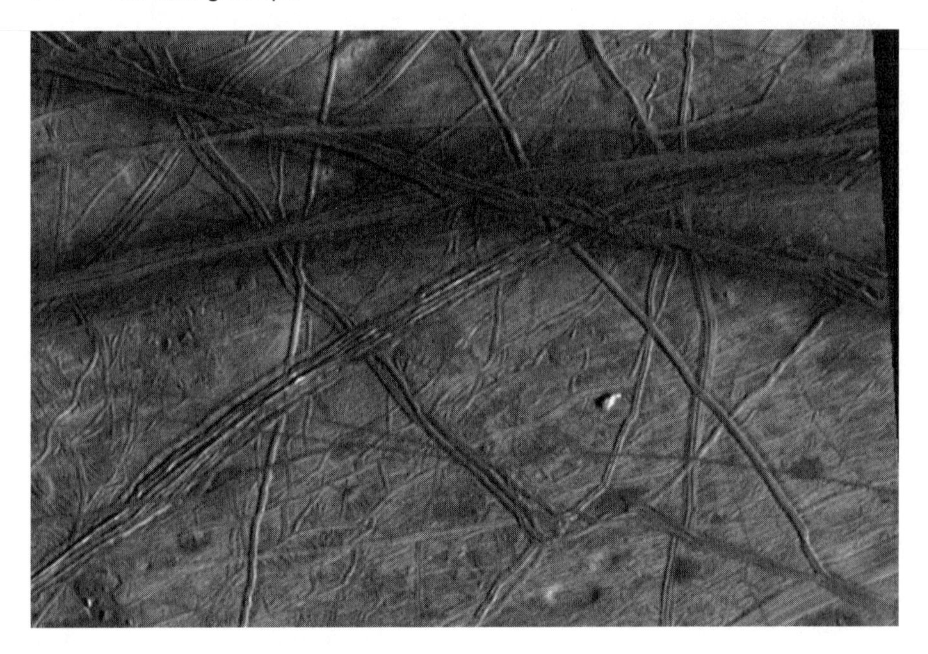

FIGURE 6.4: The intersection of the triple bands Udaeus and Minos imaged at about 200 m/pixel during Galileo's orbit E15. Compare this version with the early low-resolution view from orbit G1 (**Color Plate 2**). Udaeus and Minos are typical global-scale "triple-bands" seen at this resolution as complexes of double ridges, with dark margins extending several kilometers outward from their flanks. This region is very rich in cross-cutting ridges, which allows the sequence of their formation to be inferred.

Once Alyssa made the measurements and sorted it all out, she found that putting the features into a time sequence required that the direction of crack formation must have spun at least three times around over the time that these cracks formed.

The sequence of crack orientations that we had inferred several years earlier from the low-resolution G1 data had spanned only a few tens of degrees of azimuth variation, apparently only a small fraction of one nonsynchronous rotation period. On close inspection, the new results seem to pull the rug out from under the older interpretation. We took a look at the lineaments that we had identified in the G1 study and considered where they fit within the more complete sequence that we had derived from the E15 images. We found that several new lineaments fit between the G1 lineaments in the formation sequence. Even though the lineaments identified by Paul seemed to lie within a fairly narrow range of orientations, the new results showed that they did *not* all form during a single cycle of nonsynchronous rotation. Evidently, it was merely a matter of chance that the selection of identifiable lineaments in the G1 images had appeared to represent a smooth continuous change in azimuth over a few tens of degrees.

We concluded that there is no evidence in the cross-cutting record described that supports nonsynchronous rotation. Alyssa investigated several other regions of Europa using the E15 images and others from later orbits, all with the same result: Although there is a rich record of cross-cutting relationships, there is no compelling evidence in that record for nonsynchronous rotation.

That result does not mean Europa rotates synchronously, only that there is not a smooth uniform variation with lots of cracks formed in a single rotation. It is perfectly consistent with our earlier results based on the cycloids in the Voyager image, which had indicated that only a couple of cracks could form in any region per rotation period. In fact, with cracks forming so infrequently, we would expect no record of recognizable azimuthal variation, even if Europa has been rotating nonsynchronously.

When Alyssa presented these results at a couple of scientific conferences, I was surprised at how aggressively she was attacked. First, I expected that a young undergraduate would have been treated with at least some sensitivity. Second, the news she was bringing was that my group's own earlier interpretations were now in question. We were comfortable with that result, so why should anyone else get upset about it? The problem was that our earlier interpretation had been accepted into the canonical set of Galileo discoveries. It had been used by others as the basis of their own geological interpretations, and with that investment they had taken ownership of the idea. Now, in reversing our earlier result, we were pulling the rug out from under them. My approach to science has been to follow the evidence where it takes us, which is good science, but bad politics. Ironically, in this case, the part of the canon we were questioning had been our own discovery.

While our 1998 interpretation of the G1 data was reasonable given the limitations of the G1 data, the better images from orbit E15 showed us that the case for nonsynchronous rotation from lineament azimuths is hardly compelling. Similarly, continuing investigation of the large set of data acquired over the entire Galileo mission can provide deeper understanding, and can correct misconceptions based on the earlier limited data. It seems likely that Europa *does* rotate nonsynchronously, but not based on the evidence we perceived at first.

Images already in hand are likely to provide greater insight regarding many of the issues under discussion regarding Europa. We need to exploit all that information, and follow its implications. If we change our minds about what Europa is like, it will not be a sign of failure, but of success in pushing forward. To a large extent, the images and other data from the Galileo mission remain to be explored and interpreted fully. That work needs to be done if we are to justify the effort that went into the Galileo mission, and to prepare for subsequent spacecraft exploration of Europa.

As invested as I am in modeling of Europan tectonics in terms of the effects of tides and of nonsynchronous rotation, I am comfortable with the possibility that we may have it all wrong, or that things may be much more complicated than we have assumed. In various areas there seem to be favored orientations for cracking. Perhaps some preexisting weakness in the ice is a dominant effect, hiding any record of rotation. Or maybe the entire global crust slips around occasionally relative to the spin axis. In **Chapter 9** we shall see evidence, developed by Alyssa, that such "polar wander" did occur. Such slipping of the whole crust around the planet could have happened repeatedly, either at random or in some systematic way that we do not yet understand. In that case, it may be that a dominant driver of tectonic cracking is the stretching of the crust that occurs as it re-conforms to the tidally elongated figure of Europa. From the point of view of computing tidal stress, polar wander would be a sort of nonsynchronous rotation in arbitrary, or at least unsystematic, directions. If it happens suddenly, tremendous stress would ensue, possibly dominating the tectonic record. We have only begun to consider the range of possibilities. Whatever it is, the story has been written in the tectonic record, and is there to be read.

In the following chapters, we will consider the various ways that cracks manifest themselves in the visible record of Europa's surface evolution. The most common sign of an underlying crack is the ubiquitous double ridge. We have already seen that ridge formation has not only been one of the dominant resurfacing processes, but that it also has the potential for providing a constraint on the rate of nonsynchronous rotation.

Building Ridges

Ridges appear in nearly every image of Europa's surface. In **Chapter 2** we saw beautiful examples of simple double ridges, of global-scale complexes of multiple entwined ridges, and of terrain packed densely with criss-crossing ridges. Even within chaotic terrain, the rafts often display portions of earlier, ridged terrain that predated the chaos. We saw that the common denominator of ridges, in all their diversity, is the double ridge. Of the thousands of Europan ridges that I have surveyed, my favorite is shown in the spectacular high-resolution view in **Fig. 7.1**.

This view is so close-up that bumps as small as a house are visible. The view is at an angle, like looking sideways out the window of an airplane. A double ridge, curving down from top-center, rides over other ridges coming from the upper left. The double ridge is sliced off, revealing its interior, just as a roadcut created during highway construction reveals the layers of different materials below the surface. The terrain in the foreground is probably chaos, and most likely it was during its formation that the ridges were cut off and the interiors were revealed. What makes this image so special is not only that the view is amazingly close-up, but that it allows us to look inside the ridges.

Another reason I like this picture so much is that it was hidden away for a time, until my student Greg Hoppa dug it up and started showing it around. Even members of the imaging team were surprised that it existed. Naturally, I like to call it The Picture the US Government Did Not Want You to See.

If you are a conspiracy buff, consider this. The press-release version of the single highest-resolution image of Europa shows none of the "roadcut" area. What is more, the imaging team's catalog, an atlas that documented all the areas that had been photographed by the Galileo camera, omitted the picture sequence from which this extraordinary photo was taken. In principle, all Galileo images are available to the public, but such access is not useful if you do not know that an image exists. People rely on publications and press releases and catalogs to learn what is available. It did seem strange, then, that information about this image, and especially its roadcut portion, was

FIGURE 7.1: A portion of the highest-resolution image of Europa ever taken, and one of the most extraordinary. The resolution is 6 m/pixel. The area shown here is about 2.4 km across in an oblique view—from above and to one side. Double ridges curve down from the top of the image. In the foreground is chaotic terrain. The chaos has cut off some ridges, revealing their interiors.

especially hard to find. The power brokers of the Galileo imaging team, who did control public-information releases and catalog preparation, might prefer that this image remain obscure: Its detail supports the interpretation of ridge formation described in this chapter.

But, was the imaging team, or for that matter anyone else, really hiding this picture? Was this crucial piece of evidence being suppressed? Actually, it turns out that the contents of **Fig. 7.1** were cut off from the rest of the highest-resolution image when the image was downloaded from the spacecraft's tape recorder and radioed back to Earth in two separate batches. For bookkeeping purposes, the extra file with the road cuts was assigned a separate unplanned image-catalog number, and so promptly forgotten.

In one sense, this confusion about the lost image resulted from the failure of Galileo's high-gain antenna to open properly (as described in **Chapter 9**), which severely reduced the rate of data transmissions from the spacecraft. As a result, complicated strategies had to be devised to store images on the archaic tape recorder and then radio the data to Earth piecemeal, over time. Not surprisingly a few items got mislaid. In a deeper sense, then, the cause of the missing image material may go back to the space shuttle *Challenger* disaster, which had delayed Galileo, necessitating the extra transcontinental road trips (described in **Chapter 3**), which in turn shook up the high-gain antenna mechanism, perhaps causing it to fail years later in space. But, however far back one wants to trace the cause of the confusion, the image remained ignored until Greg, poking around for revealing images of double ridges, brought it back to light.

How could the ubiquitous double ridges have formed? Voyager had revealed mobile plates of surface ice, and post-Voyager theorizing had shown that these plates might have been moving over a liquid ocean. As Galileo approached Jupiter, we began thinking that the surface of Europa might more resemble the Arctic ice cap than anything else on Earth. Ron Greeley, the imaging team geologist from Arizona State University, had engaged an expert on the physical structure of Arctic ice to advise on what we might find. The consultant, Max Coon, briefed the imaging team on Arctic ice processes and structures, in anticipation of what we might soon see in the Galileo images. Coon pointed out that ridges in the Arctic can form as large plates of ice press together, suggesting that similar processes might be at work on Europa.

Europa's diurnal tides, combined with the forces described in Max Coon's Arctic analogy, suggested a systematic ridge-building process. Consider how a new crack forms by tidal stress. Suppose, at that site, diurnal tides change the stresses periodically during every orbit, but not quite enough to crack the ice. At the same time, the tidal bulges slowly but continually move around the body, as nonsynchronous rotation changes the relative direction of Jupiter. The gradually increasing stress due to non-synchronous rotation is superimposed on the diurnal stress. Then, suddenly one Europan day, as it increases toward its daily peak, the stress exceeds the strength of the ice, and the ice cracks.

Now, even after the stress had been relieved by cracking, the body of Europa would continue to undergo its daily distortion. (Recall, one day on Europa is about 3 1/2 Earth days long.) As long as the crack did not anneal by refreezing, it would be regularly worked, opening and closing on a daily basis. In other words, the diurnal tidal stretching of Europa would be accommodated by periodic opening of the cracks, rather than stretching of the elastic layer of ice.

Ridge-building, as I envisioned it, occurs over those many small cycles (**Fig. 7.2**), in a process that can be broken into four stages that occur each Europan day:

(a) When the daily crack opening occurs, liquid water from the ocean below rushes up to fill the gap. The ice crust is like a giant raft floating on the ocean, so the rising water can go as far up as the float line. That level is defined by the buoyancy of the ice, and it is about 90 percent of the way up from the ocean, because ice is about 10 percent less dense than the liquid water in which it floats. Ice skaters on northern lakes are familiar with the effect: Water flows up to the float line, but not to the surface, so that when it freezes, it leaves a crack deep enough to catch a skate blade. Similarly, in crack openings in the Arctic (called "leads", because they can lead an Arctic ship through the ice), new ice forms at the surface of the newly exposed water. On Europa, the freezing process must be very dynamic because the water would be frigid and boiling at the same

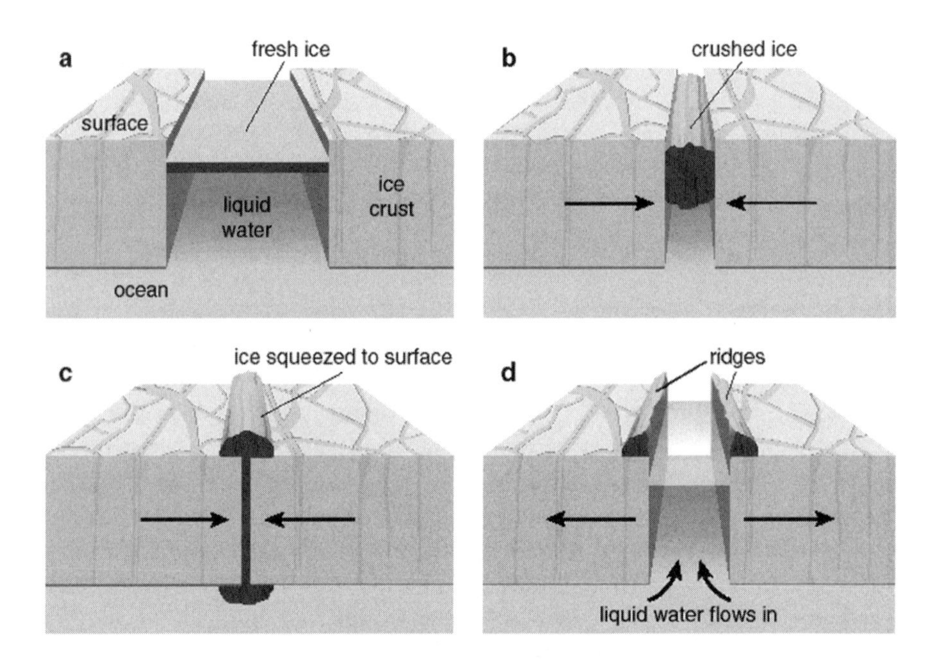

FIGURE 7.2: A model for ridge formation by diurnal working of a pre-existing crack. This four-stage cycle repeats each Europan day (3 1/2 Earth days).

time—because it is opening into a vacuum. The lead ice might well be very weak, porous, and irregular. Ray Reynolds, a NASA geophysicist who along with Stan Peale had predicted the tidal heating in the Galilean satellites, estimated during the early post-Voyager years that such newly exposed water would freeze at the surface within a few hours, to a thickness of about half a meter. After that, the freezing is somewhat slower, limited by the rate of conduction of heat from the bottom of the new lead ice out to the surface.

(b) After a few hours, as tidal stress reverses, the crack begins to close. The lead ice is crushed in the press and, as it is, some new liquid is exposed. This means that additional freezing and boiling take place, increasing the amount of ice near the top of the crack. Visually, the jumble of ice being squeezed between the walls of the crack reminds me of the scene in the first Star Wars movie in which the walls of the garbage disposer inexorably close together on the jumble of floating debris. And whether it is styrofoam and galactic rebels on the Death Star, or crushed ice on Europa, the raft of junk is squeezed thicker and thicker as the wall closes in. On Europa, at the same time as the ice is crushed and jumbled, more liquid water is exposed and freezes, adding to the volume of crushed ice. The raft of jumbled ice must become very thick as the crack closes.

(c) Then, as the crack approaches closure, the top of the now-thick raft (perhaps a slushy slurry mixed with cold boiling water) is squeezed out onto the surface, building a ridge. Simultaneously, a similar amount of material would be squeezed in the opposite direction, into the liquid ocean down below the ice, presumably to remelt quickly in the relatively warm water.

(d) In the next diurnal cycle, as a result of a new round of tidal action, the crack reopens, with some ice debris falling back into the crack, but also leaving debris in ridges at the surface along both sides of the crack. The process begins to repeat itself as ocean water rushes up into the crack.

We can roughly quantify this model using what we know about the scale of the diurnal tides. Having already commented about the ways in which geologists deal with numbers, let me demonstrate a peculiar technique favored by astrophysicists. It's called the order-of-magnitude estimate. The diurnal part of the tidal elongation of Europa's entire body is about 60 meters. (Remember from **Chapter 4** that the height of the tidal bulge at each end is about 30 meters.) This stretching is accommodated over the 3100-km diameter of Europa. So the fractional distortion (which we call the "strain") must be ~2 meters per 100 km.

The little symbol ~ can be thought of as an astrophysicist's equal sign. It means "of the order of magnitude of," or "within a factor of a few," or "roughly equal to," or perhaps even "very roughly equal to." So when I confidently tell you the crack should open by ~2 meters, I mean maybe 50 cm, or 6 meters, or something like that. This approach to numbers works well in the extreme size-scales dealt with in astrophysics, and can even be applied judiciously in geology. But do not try this at home, especially in your checkbook.

Even with its uncertainty, the tidal *strain* value is reasonably well defined compared with tidal *stress* (which represents the forces acting in the ice). The strain value, ~2m/100km, does not tell us anything about the properties of the ice, just how much the ice is stretched by tides on Europa. By contrast, the amount of elastic *stress* that this tidal distortion exerts in the ice is another story. It does depend on a key property—the elasticity—of the ice. For the elasticity of typical ice, and for this typical tidal strain, the stress is ~15 pounds per square inch (or ~10^5 Pa in metric units). By chance, that value is about the same as the pressure of the Earth's atmosphere, which is completely irelevant on Europa. More significantly, that value happens roughly to match the tensile strength of ice. In other words, if you stress ice that much it may crack. Once the ice shell cracks, it still must periodically change its global elongation to accommodate to the diurnally changing tidal shape of Europa This strain will open and then close the cracks over the course of each Europan day.

How much each crack opens and closes would depend on crack spacing. Ridges can be seen packed densely together over much of Europa. But their cross-cutting shows that they represent a wide range of ages, with the freshest-looking ridges tens to hundreds of kilometers apart. If we assume the spacing between active cracks at any one time is typically about 100 km, the tidal strain calculation above tells us that open cracks would be a couple of meters wide.

Now we have everything we need to calculate how fast ridges grow, with a precision that might satisfy an astrophysicist, if not your accountant. Every Europan day, fresh ice freezes in the active crack in an amount that is about a meter thick and a meter wide. We do not really understand all the freezing and squeezing in detail, so let's be conservative and assume the extrusion to the surface is not very efficient: If only a few percent of the crushed ice is squeezed out each day, enough volume would be cumulatively deposited on both sides of the crack to make the typical larger Europan ridges (~1 km wide and ~100 m high) in only about 30,000 years. The more common, smaller ridges would build up over a much shorter span of time. This calculation shows, then, that even though each daily cycle has only a small effect, cumulatively the cycles have played a significant role in Europan geology.

In the course of discussing our ridge-building theories with various members of the media, we supplied sketches of the process to illustrators and other graphic artists. The results included some amazing pictures. The drawing that originally appeared in *American Scientist* magazine (**Fig. 7.2**) was very helpful in conveying the basic ideas contained in our four-step conception of how ridge-building happens. NHK television in Japan, taking another approach, developed a stunning animation of how the process might actually look. (Two frames of their animation are shown in **Fig. 7.3.**). Artwork this compelling makes it easy to forget that ridge-building as we have described it is strictly theoretical, and has never actually been observed. Too often slick artwork can make even a weak theory seem great. In this case, I like to think that these artists' conceptions represent what might plausibly be happening on Europa.

Still, there is a problem with both of these illustrations (**Figs. 7.2** and **7.3**) as there must be with any illustration of ridge-building dynamics: In order to show the flow of water and the rafting of the ice, the illustrator or animator needed to make the cracks appear much wider than they really are. Remember, the ice must be at least a few kilometers thick, but the cracks are probably no more than a few meters wide. If these sketches were done to scale, they would show the thickness of the ice 1000 times thicker than the crack is wide.

Beautiful, suggestive illustration is one thing, compelling scientific argument is another. Does my story still hold up, given that 1000-to-1 geometry? Such a thin sheet of liquid would be susceptible to freezing as it flowed past the cold ice around it. On the other hand, each day's new charge of liquid would warm the adjacent ice somewhat. Both of these are reasonable views, and both are open to legitimate questions. Other outstanding issues involve details of how each day's newly-extruded crushed ice gets added to the pile on each side. The daily upwelling probably would not rise high enough to flow up and over the tops of the growing ridges. More likely, it pushes out on the bases of the ridges, making room for itself as it forces the previously extruded ice further from the crack. Some of it would probably fall back into the crack, to be re-extruded the next day.

Whatever we may learn as we consider these details more carefully, this explanation for ridge formation seems most consistent with their appearance as piles of loose, crushed material (as seen in **Figs. 2.8 and 7.1**, for example). Most compelling is how naturally the model follows from the periodic opening and closing that we expected from the diurnal tide. William of Ockham was our guide.

***O *

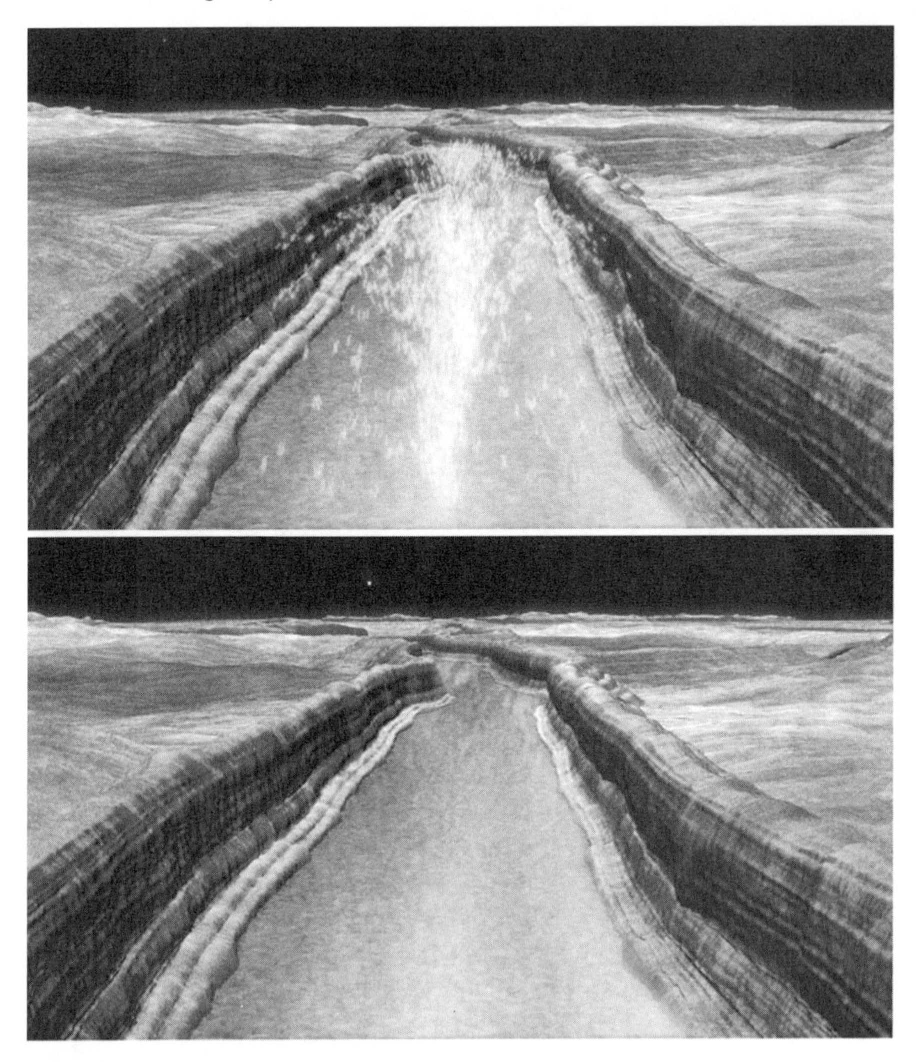

FIGURE 7.3: *Top:* Newly exposed water in an opening crack (Step [d] in Fig.7.2) is boiling as it freezes in this illustration from the NHK (Japan) TV production "Space Millennium." *Bottom:* A couple of hours after the liquid is exposed, a thin layer of ice has formed (Step [a] in Fig.7.2).

Another model of ridge formation was developed shortly after ours by two graduate students who were working with Jay Melosh, another professor in my department. Elizabeth (Zibi) Turtle and Cynthia Phillips worked with computer software that Jay had developed to simulate tectonic processes. They considered what would happen if two crustal plates squeezed together as a crack closed, with a bit of extra material wedged into the crack. They noted that such a condition might occur in the likely event that the raft of crushed ice from a frozen oceanic intrusion were not completely squeezed out as the crack was sealed. The simulations showed that a modest lip of upwardly pushed ice would form on either side of the crack. This effect almost certainly contributes to the development of ridges. My guess would be that each ridge consists of a pile of squeezed-out, crushed ice on top of a raised lip generated by Zibi and Cynthia's process.

When Zibi and Cynthia prepared a presentation summarizing their findings at a meeting of the American Geophysical Union, they created a small storm within the imaging team. Team leader Mike Belton decreed, under pressure from one of the powerful team members, that they would not be allowed to publish because they were not members of the imaging team. The rationale was that imaging team members had first rights to the data because of our decades-long work in support of the mission. Although NASA policy has now changed toward more open access to mission data, the Galileo policy had been standard for most large space missions up to that time. Numerous pictures and descriptions of double ridges had already been published, going back to Voyager images. Yet the imaging team was claiming control of all publication rights on the subject. Whether this power to control publication made sense or not, whether it was fair or intellectually honest or the decent thing to do—it was an enforceable power. Most planetary scientists depend on NASA for funding or at least for information, and if the powers in control are displeased, a project or even a career can be placed in jeopardy. Consequently, most mainstream scientists try to abide by the policies of space missions, even if they are not parties to the mission. And for young graduate students especially, the prospect of making enemies of the powerful holders and distributors of information must have been disconcerting, to say the least.

Eventually we were able to arrange for Zibi and Cynthia to present their clever idea at a couple of major scientific conferences. But after their difficult experience they tabled that line of research and moved on to other things. Still, I remain convinced that the process they envisioned probably plays a role in ridge formation—as a kind of extension of the process of extrusion by diurnal tides—and that their model deserves further consideration.

By 1998, Jim Head and Bob Pappalardo were firmly committed to the canonical thick-ice model: To state the basics, their model featured convection and upwelling material within a layer of ice tens of kilometers thick. To explain ridge formation. they proposed a conceptual model based on hypothetical long, low-density blobs within the ice crust. On Earth, a blob of low density rock that rises up toward the surface is called a *diapir*, so this notion was called the linear diapir model. In this model, a crack in the cold brittle surface is periodically worked by tides, which in turn creates friction in the viscous ice just below the crack. The friction heats the deeper ice enough to create the low-density diapir. Then, as the diapir rises up below the crack, it pushes up the lips, tilting the surface upward along the sides of the crack and creating the double ridges.

One of the most impressive pieces of evidence they presented in support of their theory was the identification of triple ridges. Their example was a 6-km-long segment shown in **Fig. 7.4**. In this ridge system, the central ridge was interpreted as the top of the diapir, rising up between the two upturned lips of the crack.

That interpretation would have been reasonable if such triple ridges actually existed. But Head and Pappalardo must have forgotten to look at the rest of this ridge system, beyond the postage-stamp-sized cut-out that they used—and that's presented in this book as **Fig. 7.4**. Just looking a few kilometers beyond (as in **Fig. 7.5**) reveals a very different story. The supposed triple ridge is really part of a complex of multiple ridges, probably sets of double ridges, that have been considerably modified by overlapping one another and by an adjacent patch of chaotic terrain, just above the

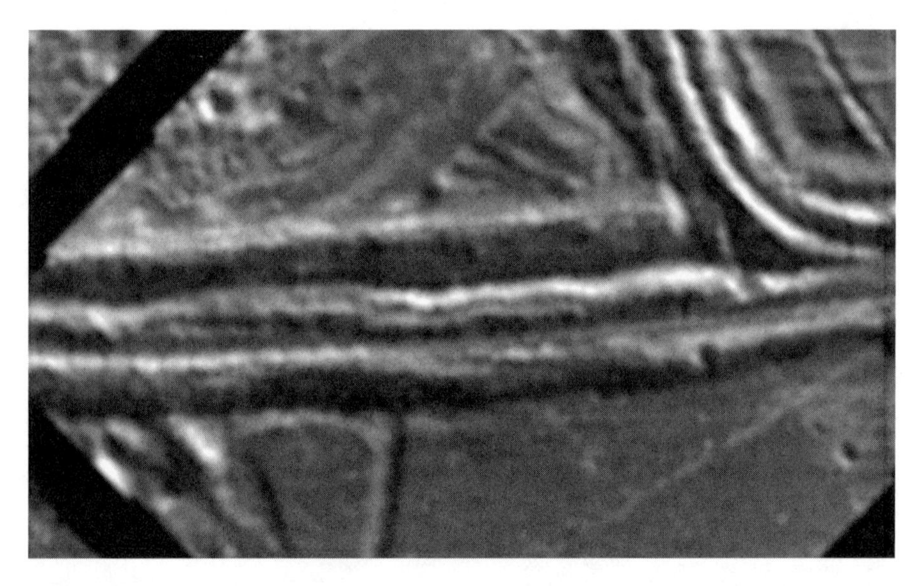

FIGURE 7.4: The archetypical example of a triple ridge, a class of feature that does not exist, as cropped and displayed by Head and Pappalardo. The portion of the ridge system shown here is about 6 km long. **Fig. 7.5** reveals the neighborhood around this tiny cut-out.

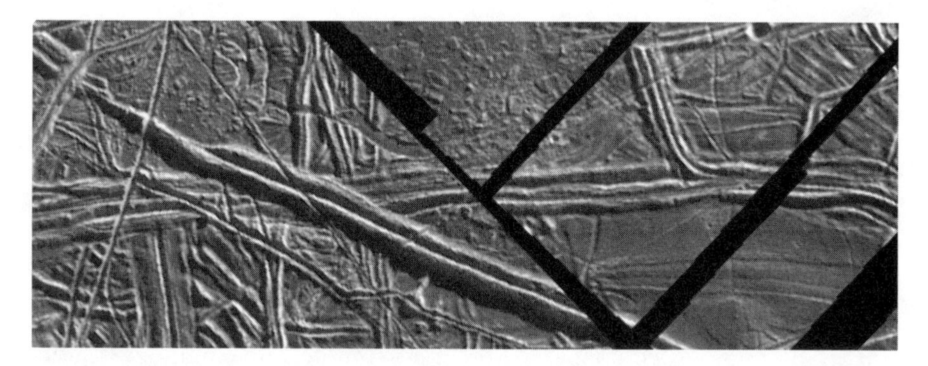

FIGURE 7.5: The terrain a few km beyond the postage-stamp cutout in **Fig. 7.4**—indicated here by the white rectangle superimposed on the image. The straight black slashes are gaps in the data acquisition.

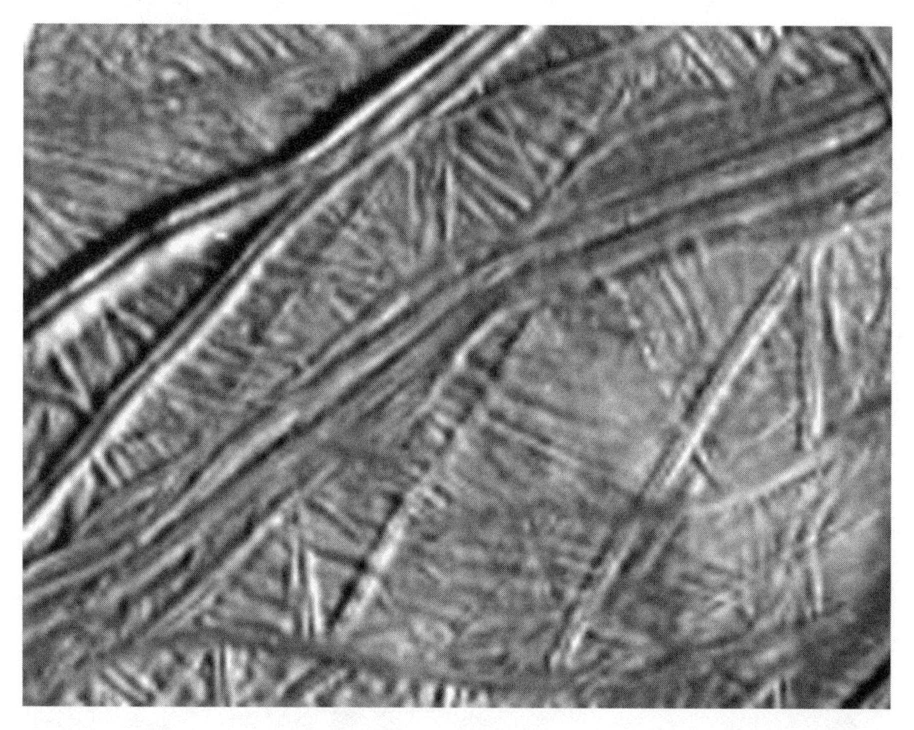

FIGURE 7.6: The complex of double ridges running from the lower left to the upper right in this cut-out was mapped as a triple ridge by Jim Head's research group at Brown University. It is no such thing. Note that this area appears in **Fig. 2.11**, and lies just northeast of Conamara.

"triple ridge" in the figure. The chaotic terrain has clearly disrupted ridges in this area. In one place, the formation of chaotic terrain disrupted one ridge of a double ridge, leaving only a single ridge for a couple of kilometers of its length. The triple-ridge cited by Head and Pappalardo appears to have been created when a multi-ridge complex was reduced to three ridges along a short part of its length due to disruption by chaos. This feature is not a characteristic product of ridge formation.

The only other example of a supposedly triple ridge was mapped by Head, Pappalardo, and their student Nicole Spaun in an area near Conamara Chaos. The feature so identified in their map, shown in **Fig. 7.6**, is clearly just a set of roughly parallel and very typical double ridges. So, the triple ridges that made the linear diapirs model seem plausible simply do not exist.

If the ridges did form by a linear diapir tilting the surface away from the crack, you might expect to find the older adjacent terrain continuing up the outward facing slopes of the ridges. Sure enough, Head and Pappalardo showed a couple of examples. However, the fact remains that of the tens of thousands of kilometers of double ridges that have been imaged, places where features on the flanks are lined up with nearby ridges are very rare. And even where the adjacent terrain does seem to continue up a ridge flank (as in **Fig. 7.7**), the appearance fits my model. As crushed ice was piled onto the growing ridges, it slid down the outer slopes. Where it was supported by the pre-existing underlying topography, raised ridges of loose material were formed on the flank. The view in **Fig. 7.7** seems consistent with the

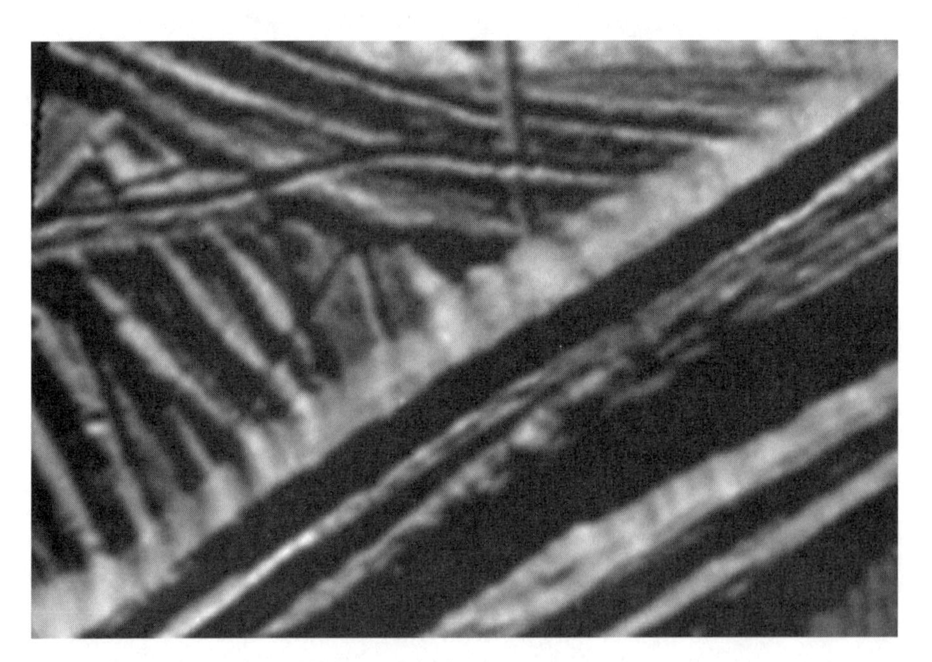

FIGURE 7.7: Near the south pole, this bright flank of a ridge (running from the upper right corner down to the lower left corner) *appears* to have extensions of older ridges running up its flank.

observation that the ridges on the flanks run perpendicular to the main ridge, rather than following the direction of the underlying older ridges. Thus even in these rare cases, the appearance of the ridges fits our model of ridge-formation by tidal extrusion of crushed ice.

The "linear diapir" model suffers from the flawed observational evidence: the lack of expected central ridges of upwelled material, and the absence of old terrain on the flanks of the ridges. Moreover, it is hard to understand how a diapir could be so uniform along many hundreds of kilometers. Consequently, the linear diapir model for ridge formation, although once a key part of the isolated-ocean story, has not been widely embraced.

A great strength of our tidal-tectonic extrusion model is that it provides a natural explanation for some of the characteristics of larger, more mature ridge systems, including secondary characteristics—downwarping, marginal cracking, multi-ridge complexes, and dark margins—that we'll discuss briefly here.

Double ridges often have fine cracks running along parallel to them on either side. You can see some next to the large double ridge in **Fig. 2.8.** Because our ridge formation model piles material on top of the surface, the brittle, elastic surface must be bent down, or "downwarped," under the weight, and it should be no surprise to find cracks along the downward bend. Similar cracking occurs along ridges loading down the Arctic ice cap here on Earth.

In some places, the downwarping of the burdened surface is obvious. A double ridge near Conamara was one of the first where this effect was recognized (**Fig. 7.8**). The geometry of the bending of a weighted elastic sheet is well understood theoretically. As soon as he noticed this ridge, my student Randy Tufts realized that it could be used to estimate the thickness of the elastic upper portion of the ice (the "lithosphere"); his calculation came to about 300 meters.

But remember Silly Putty: The faster you strain it, the more elastic it is. Randy's result applies to the part of the ice that acts elastic even during the long slow sagging under the weight of the ridge. In this case the stress is of very great duration, so only the very cold part near the surface acted elastic. For tidal stress, changing on a daily basis, the elastic part would be much thicker.

Another thing to bear in mind is that this ridge is in a special location, running right up to a very large fresh patch of chaotic terrain, Conamara Chaos. Part of the ridge in **Fig. 7.8** even lies on rafts that have broken off from the shore, although they have not been displaced much. Because

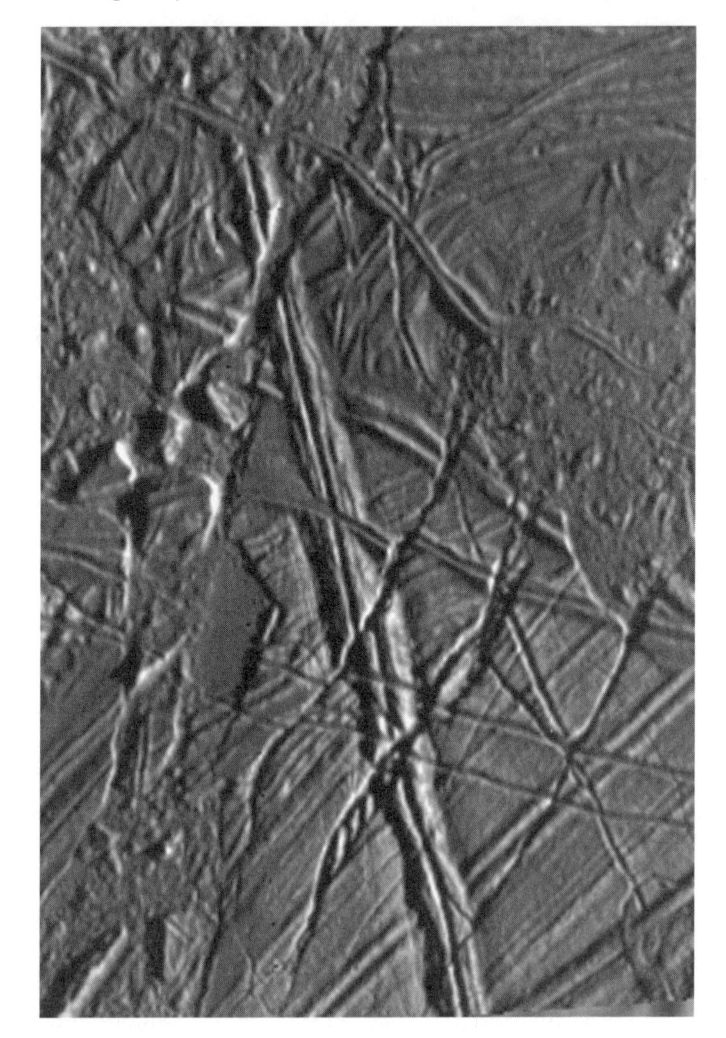

FIGURE 7.8: This double ridge runs into Conamara, where it was cut off by formation of the chaotic terrain. Shown here in a high resolution image ($\sim$30m/pixel), it also appears at the bottom center of **Fig. 2.7**. This ridge pair has weighted down the lithosphere below it, causing downwarping on either side.

Conamara, like other chaotic terrain, appears to have formed by melting from below, the elastic layer might have been temporarily thinner under this ridge.

In any case, an elastic thickness of 300 m is reasonably consistent with the expected average temperature profile for heat-conducting ice of a few kilometers' thickness. But it is also consistent with the temperature profile for convecting ice, which would increase in temperature quickly with depth in the top couple of kilometers, and then stay more or less uniformly warm down through the thick viscous convection layer.

Downwarping may have other major effects. I already noted the cracks that form on top of the downwarping. The common multi-ridged complexes may be a related phenomenon. Those large-scale ridge systems that show up in global and regional imagery as the "triple-bands" actually are complexes of roughly parallel double ridges. The large complexes that cross just north of Conamara are classic examples (**Fig. 2.7**). Such complexes may result from the "marginal cracking" that accompanies the downwarping under ridges in the following way. Once the marginal cracks are formed, they may be activated by diurnal tides, creating additional nearly parallel sets of lateral ridges.

The downwarping of the lithosphere around large ridges may also help explain the dark margins that are so common. Once the surface of the original ice crust is pushed down by the weight of the ridge to the level of the water line, liquid water might ooze over the surface. Impurities carried in the oceanic water (salts, sulfur compounds, perhaps even organic substances) might darken the surface. Association of the dark margins with oceanic substances is consistent with the similar darkening and coloration at chaotic terrain, which may represent sites of melt-through from below.

Other mechanisms have been proposed for creating the dark margins, and they are consistent with the linkage of the surface to the ocean through the cracks. Early in Galileo's imaging of Europa, Ron Greeley suggested that the dark material might be sprayed out from the cracks, perhaps as a result of oceanic exposure. This idea is appealing, because it would also explain the darkening, with apparently similar substances, of the surface at chaotic terrain. Also, if the characteristic spray distance is about 10 km, as indicated by the width of the dark margins bordering "triple bands," it would also explain the 10-kilometer-wide margins that extend beyond the edges of patches of chaotic terrain, which define the ~10km lower limit to the size of the dark spots (the "lenticulae").

The main problem with this explanation for the dark margins is that the spray would be expected to darken the ridges as well as the neighboring area, yet the ridges themselves are bright, forming the light-colored central portion of the supposed triple-bands. Presumably the ridges are bright because the darkening agents were distilled out during the freezing of the lead ice that eventually was squeezed onto the ridges. Perhaps the final part of the ridge-building occurred after most of the spraying had been completed. The spray hypothesis is very interesting and definitely needs to considered more carefully. It could be an important part of what gives Europa its distinct appearance, but we just do not understand well enough how, or even whether, it could have worked.

Another idea for creating dark areas was suggested by Sarah Fagents, a volcanologist who worked for Greeley at the time. Her idea was that warming from below may have sublimated water ice off the surface, concentrating impurities and darkening the area. This model would help explain why some areas around chaotic terrain are not only darkened, but in some cases have their topography softened (this idea is discussed in more detail in **Chapter 13**). Similarly, because ridges formed where cracks have allowed oceanic water to regularly flow to the surface, heat is brought up to the ice bordering the cracks. It seems possible that the darkening along the margins of highly developed ridge systems is a consequence of such warming near the surface.

***○ *

Double ridges, the most common and ubiquitous (and one could even say the defining) feature of Europa's surface, seem to be a direct consequence of ocean water regularly reaching the surface. The evidence is that major ridge systems correlate with tidal stress, so they must have formed along cracks. And to build ridges, those cracks must penetrate to liquid water. What does this line of thinking say about the thickness of ice? We can address that question using one other fact: Tidal stresses cannot, in all likelihood, drive cracks very deep.

The reason tides cannot crack deep ice is "overburden pressure." If you dive deep under water, the weight of the water above squeezes against your body. Indeed, if you live on the surface of the Earth, even the weight of the air above you exerts a significant pressure. And rocks deep underground are squeezed by the pressure of the overburden of the rock above. In much the same way, the ice below the surface of Europa is under pressure. And, as in all of these cases, the deeper you go, the higher the pressure.

It does not matter whether the material is solid (like rock in the Earth or ice on Europa) or fluid (like water or air). Because the overburden pressure acts over a very long time, the material below responds as a fluid. The pressure is constant and it is isotropic—the same in all directions. (On Earth you can test for isotropic pressure using your ears whenever you change altitude or dive under water: The pressure is independent of how you tip your head.)

Knowing the mass of ice and the force of gravity on Europa, it is easy to calculate the overburden pressure. About 100 m down, the pressure would be about the same as the atmospheric pressure on Earth. What is more relevant, it is about the same as the tidal stress on Europa. (The tidal stress on Europa is ~100,000 Pa, or Pascals, a unit of force per unit of area, appropriate for measuring stress or pressure.)

Deeper than ~100 m the pressure pre-stresses the material so much that, when the tide stretches the ice, the most it can do is relieve the compression. It cannot induce cracking, because the tension is overwhelmed by the compressive overburden from above. Still, there may be ways that cracks can go deeper. Suppose a crack starts at the surface, where the ice is very cold and brittle and there is no overburden pressure. As the crack grows downward, extra stress would be concentrated at the base of the crack, allowing it to go deeper than our 100-m limit. Even so, it is hard to imagine the crack going deeper than a few kilometers. Several other factors could affect the depth of crack penetration. If liquid water got into the crack, for example from the walls of the crack as they are warmed by friction, it would provide internal pressure that would push outward against the walls, balancing the overburden pressure. Also, if chunks of ice fell into a crack, they might help wedge it open and drive the crack deeper.

So cracks could penetrate well below the 100-m limit set by overburden pressure. On the other hand, it is very difficult to imagine how any crack at the surface could extend down more than a few kilometers. Therefore, if our proposed mechanism for ridge formation is correct, it implies that the ice must be fairly thin, probably less than 10 km.

That result is not consistent with the canonical model of thicker convecting ice. Under the mantle of authority, a recent editorial in *Icarus*, considered the most prestigious journal in the field of planetary sciences, reduced my contributions to studies of Europa to a single item: It quoted me as arguing that the ice is less than 1 km thick. I never wrote that; in fact I never said it, either. What I and my colleagues have argued for years is that the ice is thin enough that the ocean is connected to the surface through cracks and other openings, which probably means the ice is less than 10 km thick. The authorized story distorted my views, but cannot change the scientific evidence. The canonical model of thicker, convecting ice is not consistent with the evidence for linkages of the ocean to the surface.

Mind the Gap

The crust of the Earth is made up of several, large, rigid plates of rock that slide around the surface. Most dramatic geological activity takes place at the boundaries, where plates pull apart, slide past one another, or converge. On Europa, too, the same three kinds of displacements have occurred along the cracks through the crust, which define the edges of shifting crustal plates. In the next three chapters, I describe the three types of displacement: First dilation, where cracks open and the plates of the crust on either side move apart; then strike-slip motion, where plates shear past one another; and then convergence, where plates crash together.

In the case of dilation, cracks open, plates move apart, and new surface is created between them. The classic example on Earth is the opening of the Atlantic Ocean, the separating of continents on either side, and the creation of new crust on the widening sea floor. Similar spreading has just begun along the East African rift, and its extension up through the Red Sea. On Europa, the simple double ridges that dominate much of the tectonic terrain show no signs that the cracks have widened, other than the periodic diurnal opening and closing. The extruded crushed ice has been emplaced on top of the adjacent surface rather than filling a dilating gap. However, some ridge systems do seem to be associated with dilation of cracks. My student Randy Tufts spotted the first example, shown in **Fig. 8.1a**. First, he noticed that this ridge system looked different from most others. It had a central groove, but the ridges were much broader and relatively flat compared with most double ridges. He noticed too that the patterns of features on the adjacent plates seemed discontinuous. After staring at the picture for a while, he took out a pair of scissors and snipped the broad ridges out of the picture, then moved the parts on either side together. When he did, the patterns of ridges on the adjacent plates lined up perfectly (**Fig. 8.1b**).

In effect, Randy took this image backward in time, reconstructing the

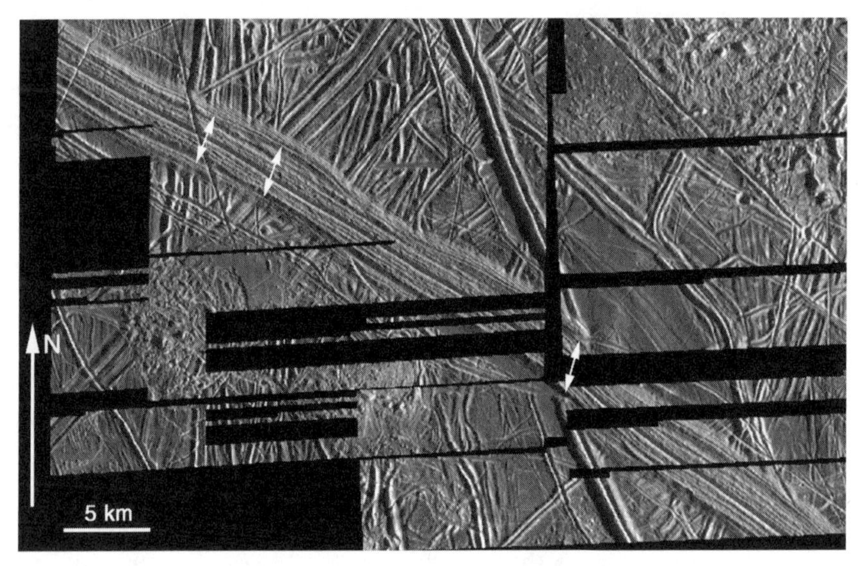

FIGURE 8.1a: A dilational ridge runs from upper left to lower right; it is recognizable because of its central groove with multiple, symmetrical ridges on both sides. The white double-headed arrows show matching points of older lineaments that were cut off and dislocated by the dilational ridge system. The black gores (empty lines and panels) in this image represent areas where we have no data. (Also note the supposed "triple ridge" to the upper right of center; it is the same as the one shown in **Figs. 7.4 and 7.5**, and is in fact a product of modification by the chaos to its upper right.)

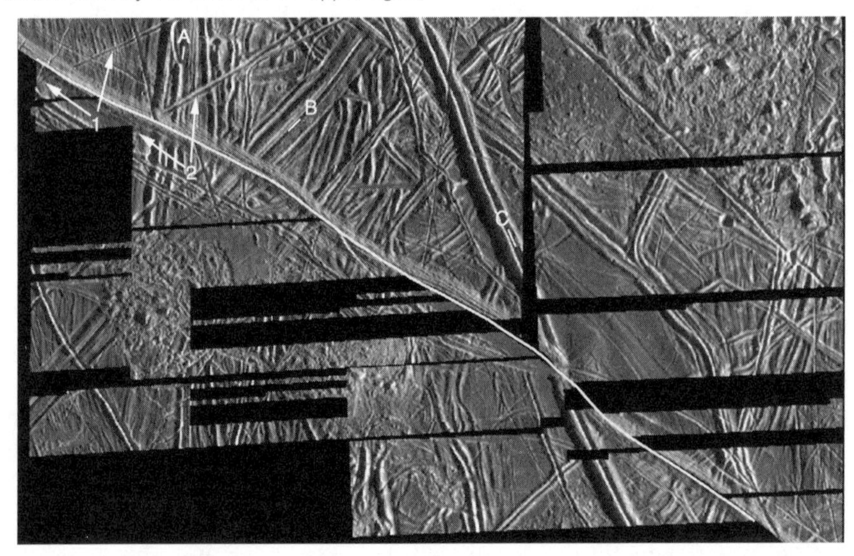

FIGURE 8.1b: The region in part (a) has been reconstructed by cutting out the dilational ridge, and moving together the sections of terrain from either side. Older linear features are now restored to their original straight, continuous configurations. (Markings and reconstruction by Randy Tufts.)

FIGURE 8.2: Ridge formation during dilation builds multiple, symmetrical ridges. Here, I have also drawn parallel cracks that cut down through the terrain adjacent to the ridges. Such cracks are often found on Europa running next to ridges, probably due to the weight of the ridge bending down the crust a little bit.

appearance of the surface before the crack had dilated. And he showed that the broad ridge system had filled in the gap.

Most dilational ridge systems look similar to **Fig. 8.1a.** Like most double ridges, there is a groove down the center, but where dilation is involved, the ridges on both sides are much broader. These ridges often have fine grooves running along their broad tops, forming lots of "mini-ridges." These structures are generally symmetrical about the central valley.

This typical form makes sense if we think about the ridge-building process (**Chapter 7**) and how it would be affected if a crack were slowly spreading. Suppose ridge-building had begun as a result of the diurnal opening and closing of the crack. Now, if a very slow, gradual separation is superimposed on that daily cycle, the two ridges will slowly move apart. But each day, new crushed ice will keep the gap filled, and the crushed ice slurry will be squeezed up between the original ridges, gradually producing the symmetrical ridge sets with the crack at their center (**Fig. 8.2**).

We do not know what drives crack dilation on Europa. Various processes may be in play. The same diurnal pumping that builds ridges may contribute to dilation. If the extrusion of the crushed ice is not completely efficient, some fraction of it may remain jammed between the walls within the crack. With each diurnal closure, the original walls of the crack (and the ridges on top) may be ratcheted gradually apart. For dilational ridges, where widening of the crack has clearly been accompanied by a lot of squeezing of material, this mechanism seems quite plausible.

On the other hand, it turns out that most dilation on Europa is not marked by ridges but rather by relatively smoother and broader bands, like those shown in **Fig. 8.3a**. Their dilational character can be readily demonstrated by reconstruction of the surrounding terrain, as shown, for example, in **Fig. 8.3b**. The adjacent terrain moved apart, and the sickle-shaped crack opened by tens of kilometers over a length of hundreds of kilometers. Serious amounts of new surface area formed in the creation of a single band, and such broad dilational bands are very common on Europa.

A great concentration of such dilation bands is located in the "Wedges" region, which showed up clearly in global scale images in **Chapter 2**. Several of those cracks dilated at an angle so that the bands are wedge-shaped, which led to the informal name of the region.

The Wedges region was imaged by Voyager, but the images did not show the detail that made definitive reconstructions (like that in **Fig. 8.3b**) possible. Nevertheless, the character of the displacement of the crust was recognized early on by Paul Schenk, then a graduate student at Washington University in St. Louis, and his thesis advisor Bill McKinnon. Paul's work convinced me, long before Galileo got to Jupiter, that Europa's crust was highly mobile. However, that notion was too radical for the referees judging Schenk and McKinnon's paper for the planetary-science journal *Icarus*, and publication was delayed for several years, a classic example of being too correct too soon. Also, it did not help that Schenk and McKinnon had not been on the Voyager imaging team, and did not wear the official mantle of authority, but Galileo images later confirmed their discovery many times over.

With Galileo images, reconstruction of the sickle-shaped band (**Fig. 8.3b**) and many other bands is practically seamless. Clearly, such bands formed more recently than the features around them. To me this clearly suggested again that the crust of Europa has been highly mobile up until the most recent part of its short surface history, and there is no reason to believe that such active displacement is not continuing today.

The geometry of dilating cracks is similar to the dilation at spreading zones on Earth, like the Atlantic sea floor. As the crack opens, older rock is carried further from the center, while new volcanic material rises to fill the gap. On Europa, the gap is filled by water and slush rising from below and freezing in place.

The morphology of the Sickle is typical of dilational bands, and just what you might expect from such a spreading-and-filling process. The boundaries are parallel and perfectly matched (like the coastlines of Africa and South America, but even better), allowing for reconstruction along the original crack line. The interior of the band is relatively smooth and flat, with only fine, shallow furrows parallel to its length. Often a

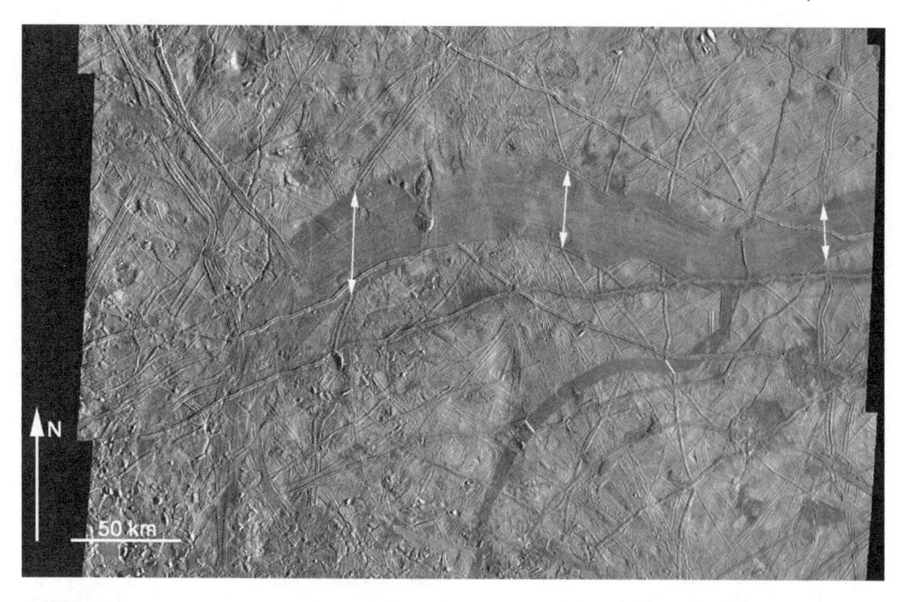

FIGURE 8.3(a): The "Sickle" is a typical dilational band, as is the curved band to its south. It lies at the northwest corner of the "Wedges" region, which we saw in its global context in Chapter 2.

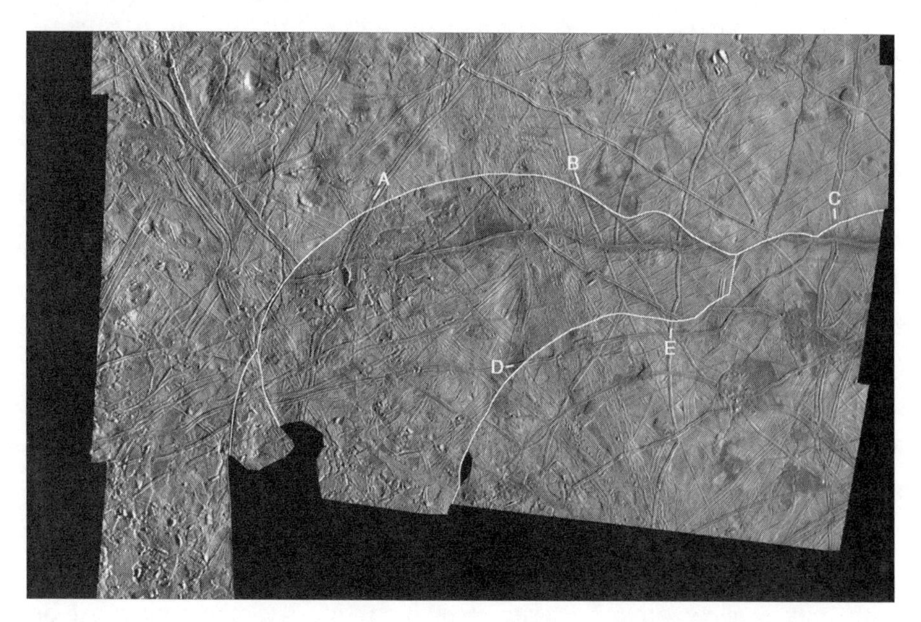

FIGURE 8.3(b): A reconstruction of the Sickle area, by Randy Tufts, shows how it looked before the bands opened up. Double-ended arrows in part (a) show "piercing points" of features that cross over the band, and that are aligned perfectly in the reconstruction.

central groove shows where the most recent spreading has been occurring. The appearance is similar to the dilational ridges (e.g., **Fig. 8.1**), except that this band is much wider and it has only the subtle furrows rather than ridges.

Why do some dilating cracks develop ridges, as in **Fig. 8.1**, while others take on the smoother character, like the Sickle? The most probable explanation is that whatever process drives the dilation, it goes faster in the case of a smoother band. If the dilation is slow, the diurnal opening and closing can pump ridge-building ice to the surface at the same time as the gradually widening crack fills with new ice. If the dilation is fast, the crack widens enough each day that the diurnal closing cycle cannot slam it shut enough to pump up much material.

As a result we see a continuum of forms, ranging from the smooth bands to the dilational ridges. They likely reflect a wide range of dilation rates relative to the diurnal oscillation. Many cases are hybrid, where during part of the dilation the rate was fast enough to emplace flat new ice, while at other times dilation was slow enough to allow ridge growth. Such ridges moved apart symmetrically as new material filled in during later dilation. The history is recorded by symmetrical ridges in otherwise flatter bands (e.g., **Fig. 8.4**).

Dilational bands, like other types of surface features on Europa, appear to have formed throughout the history of the surface (at least over the past 50 million years or so, for which we have any record). Although the most obvious examples are generally the ones that formed most recently, careful surveying can reveal older examples that have been modified by subsequent processes, such as later ridge or chaos formation. Over the course of the past few tens of millions of years, the surface has been continually renewed by tectonic processes like ridge-building and dilation, as well as by chaos-formation.

What is driving the dilation? In the case of the dilational ridges, where dilation has been slow compared with the diurnal pumping cycle, it is plausible that the plates were wedged apart by ice that was stuck in the crack. That would explain why the closing phase of the diurnal cycle would apply the pressure needed to pump up new frozen ice to the ridges. But if instead the plates are pulled apart by some regional forces on the plates on either side of the crack, there would be minimal compression even during the diurnal closure phase. That may be why most wide bands are fairly flat, with only shallow furrows rather than high ridges. Something must be *pulling* those plates apart, rather than pushing them apart from within the crack.

Evidence that dilational bands are created by such broad regional forces

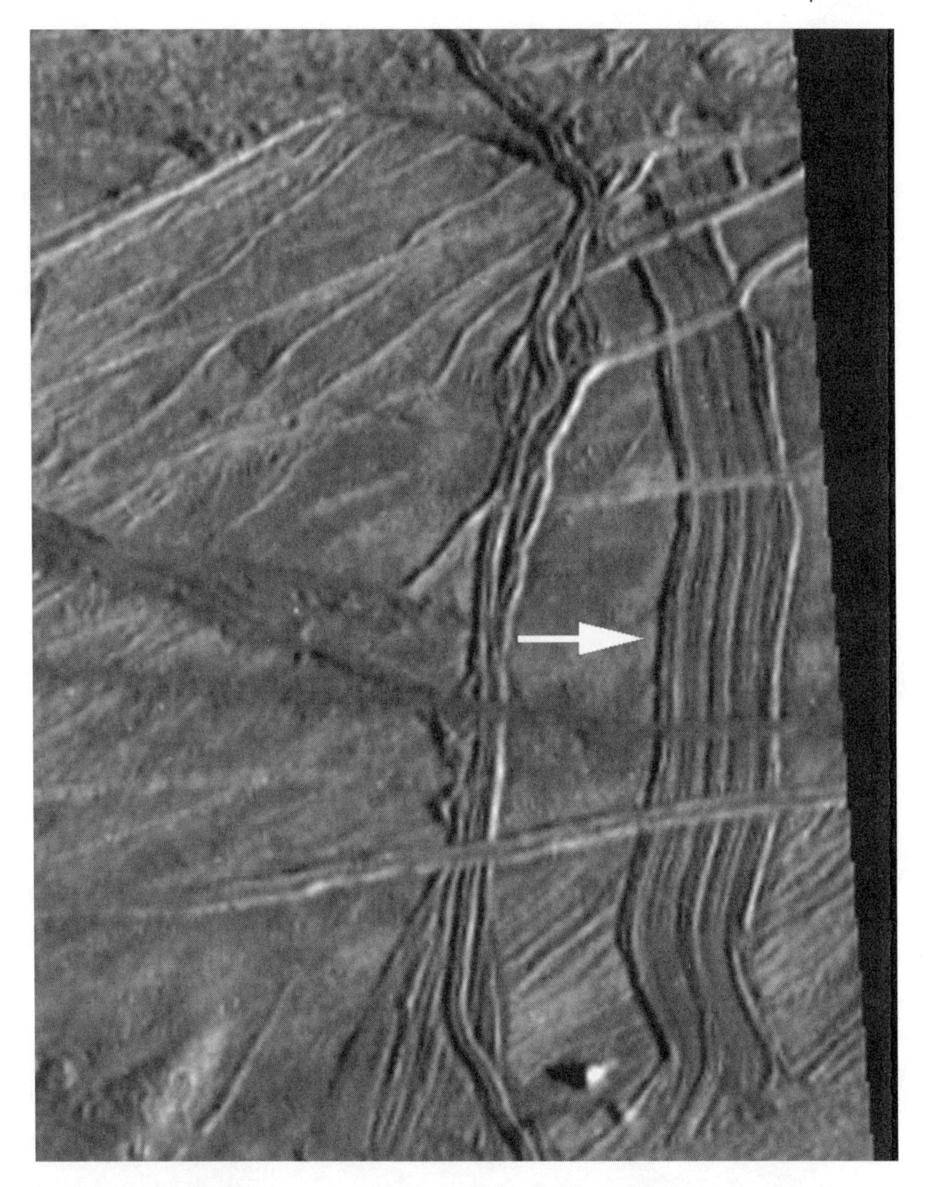

FIGURE 8.4: Hybrid bands, where dilation rate has varied during formation, producing ridges and furrows symmetrically arrayed about the centerline.

comes from my favorite example, shown in **Fig. 8.5a**. The site is a couple of hundred kilometers east of the Sickle. (For fun, you can pick out these bands in **Fig. 6.1** or near the middle of **Fig. 2.3c**, and see that the two are both extensions of the Sickle. They are slightly darker than the older surrounding terrain, including the terrain that lies between them.)

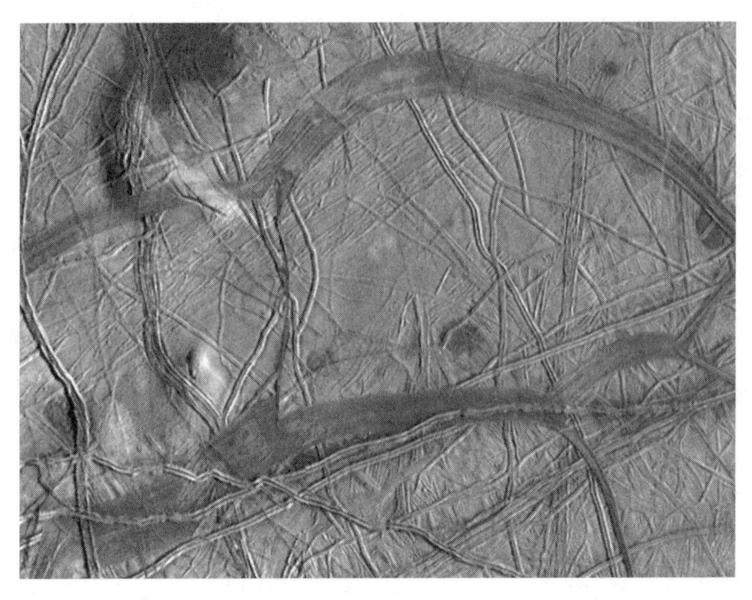

FIGURE 8.5a: A pair of parallel dilation bands, which are located east of the Sickle (Fig. 8.3). A north-south running "strike-slip" (i.e. shear) fault connects the dilation bands. The width of this image is about 180 km.

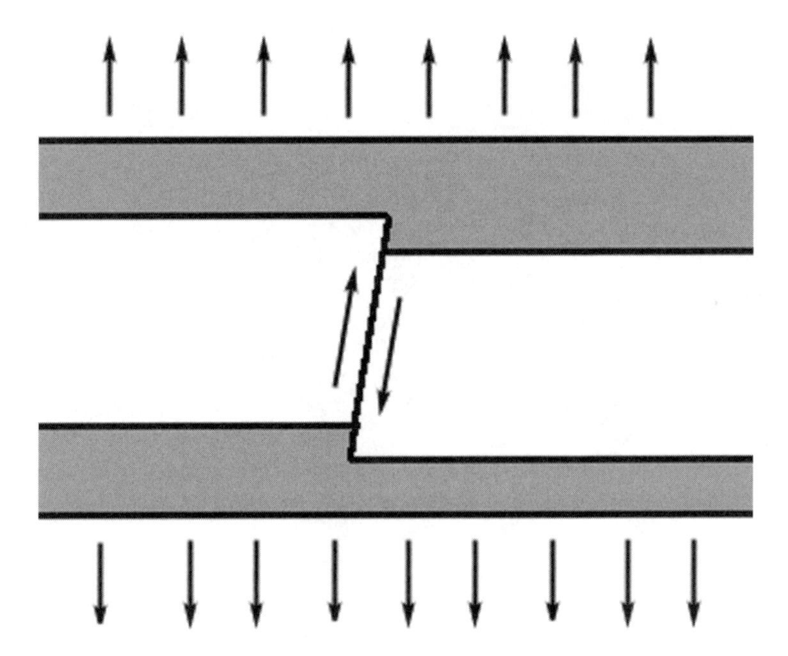

FIGURE 8.5b: A schematic of the dilational displacement that created the arrangement of crustal plates in Fig. 8.5a. A north-south running "strike-slip" (i.e. shear) fault connects the dilation bands.

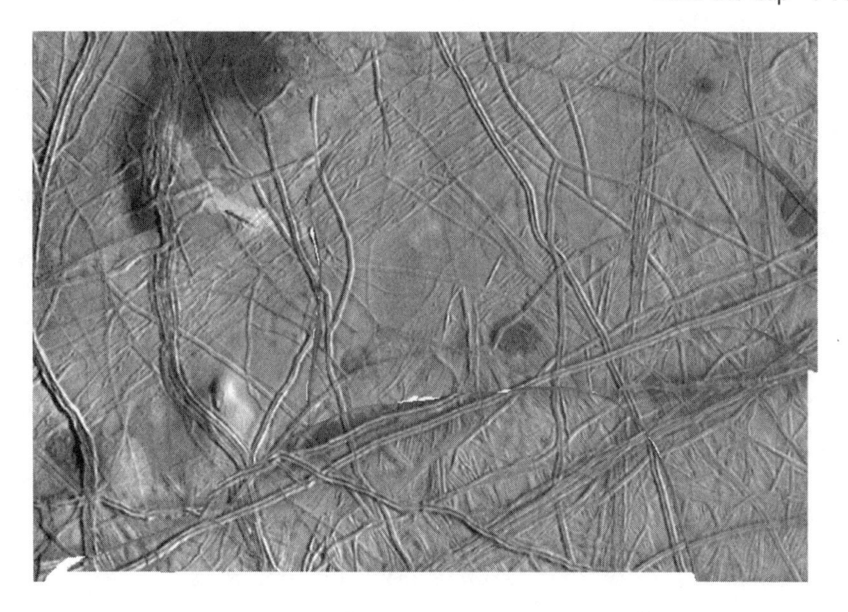

FIGURE 8.5c: Reconstruction of the band complex shown in (a), by yours truly.

I love these bands for two reasons: (1) the geometry is rather interesting and (2) I was the person who figured it out. It is my baby. I noticed that the terrain between the two bands is divided by a shear displacement. I sketched this geometry in **Fig. 8.5b**. My reconstruction, obtained by re-closing the bands and sliding back the sheared part, is in **Fig. 8.5c**. Whatever opened these cracks must have been pulling this whole region apart, not each band separately.

We can only speculate regarding the source of the force. A good guess would be oceanic currents. In that case, the ice must be fairly easy to displace. The implication would be that the ice must be fairly thin. If it were too thick, movement would be resisted by the strength of the ice in the surrounding region.

Cracks open as plates move apart, but what happens in the direction that the plates are moved? Another way to think about that problem is in terms of the surface area budget. If cracks have dilated as plates have pulled apart, then new surface area has been created. In order to balance the gain, surface area must be removed someplace else. A planet with too much crust for its size would not be a pretty sight.

For years after dilation had become well-known, the mystery of the surface area budget nagged us. Most of the other known resurfacing processes, such as ridge-building, do not affect the overall area budget. They replace just as much surface as they destroy, because the new material is placed on top of the old.

We do understand the surface-area budget on Earth. The creation of new area as plates move apart is balanced elsewhere by processes that leave very distinct, recognizable structures. Where one tectonic plate (usually relatively thin undersea crust) slides down under another we see arcuate volcanic mountains or islands, sea-floor trenches, and plenty of earthquakes (like Japan or Indonesia, for example). Where thick continental plates have crashed together, mountains have been squeezed up. The Himalayas grew this way when India collided with Asia.

Nothing quite like those familiar signs of plate convergence was evident on Europa, yet the excess area had to disappearing somewhere. Only after a few more years of studying large plate motions, described in the next chapters, did we stumble upon an answer.

Dilation is a major process for rearranging the surface and creating lots of new surface area. The forces that drive it may include ice jammed in working cracks or ocean currents, as I suggested above, but we really do not know. We have learned several things from dilation, however:

1. The driving forces operate over large regions.
2. The crust is very mobile, able to slide readily over a slippery layer, which is presumably the ocean.
3. The cracks that dilate into bands must penetrate to the ocean.
4. Given that the cracks cannot have penetrated down more than a few kilometers, the ice must be fairly thin.

Strike-Slip

Plates of crust that make up the outermost shell of Europa or Earth do not only move apart (as discussed in the previous chapter) or together (to be discussed in the next chapter). They also can slide past one another—a phenomenon I describe in some detail in this chapter. On Earth, for example, the coastal area of southern California scrapes northward past the rest of North America along the San Andreas Fault. This sort of "shear displacement," or sliding movement along cracks, is very common on Europa.

In geological parlance, the shearing of one plate past one another is called "strike-slip" displacement. "Strike" refers to the direction of a fault across the landscape. A pilot following along a fault line would call this direction the "heading." So "strike–slip" means that the displacement (the "slip") is along the strike. Strike-slip motion is *parallel* to the strike of the crack. In contrast, when a crack dilates, the displacement is *perpendicular* to the strike.

The first identification of displacement that was predominantly strike-slip was made by my student Randy Tufts in 1996, using 17-year-old Voyager images of a fault called Astypalaea. Following that discovery, with the other members of my research group, we found that strike-slip displacement is common and widely distributed on Europa. We realized that it was probably driven by diurnal tides, in a process we dubbed "tidal walking." Combining the results of our survey of observed displacements with the theory of tidal walking, we found that Europa's shell may have slipped around the globe as a whole. Moreover, our investigations of strike-slip may have solved the mystery of the surface area budget, introduced near the end of the previous chapter, by revealing sites of surface convergence, the subject of the next chapter. All of these disparate implications developed from Randy Tufts' discovery of strike-slip displacement at Astypalaea—a story that deserves a detailed telling.

Randy had found strike-slip at Astypalaea on one of the handful of

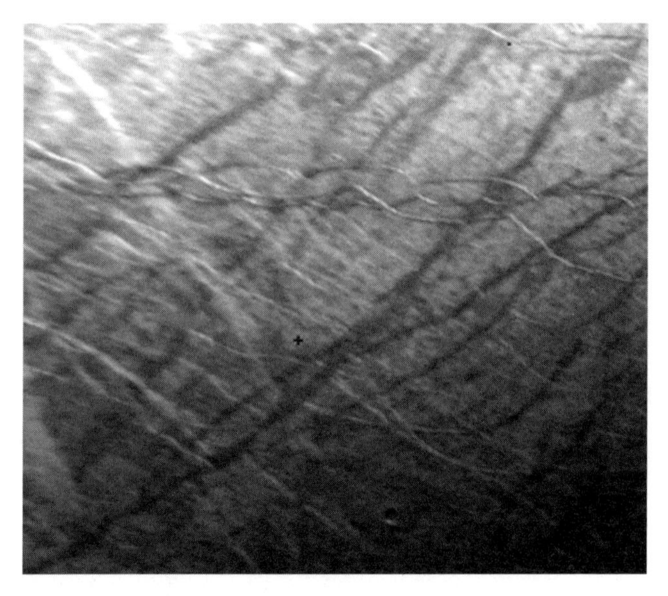

FIGURE 9.1a: Astypalaea Linea is the dark linear feature running from the lower left to the upper right in this image from the Voyager spacecraft, which has been reprojected to show the geometry as if viewed from straight overhead. The south pole is just below the area shown. Note the dark parallelogram at the upper right (northern) end of Astypalaea, and the various widenings and narrowings along the full length of the fault.

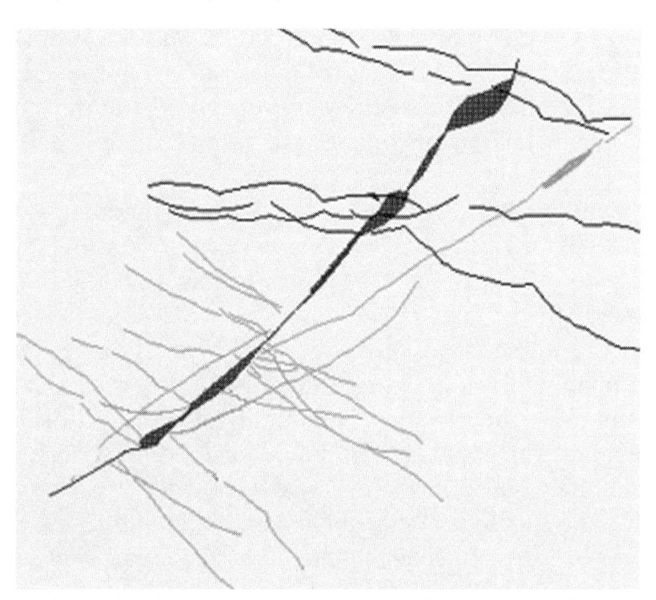

FIGURE 9.1b: A sketch of the key tectonic features in **Fig. 9.1a**: The varying width of Astypalaea Linea itself, the more recent cycloid-shaped (chains of arcs) ridges, and the wispy, old, white lineaments that end at Astypalaea and were evidently cut-off by it (based on a diagram by R. Tufts).

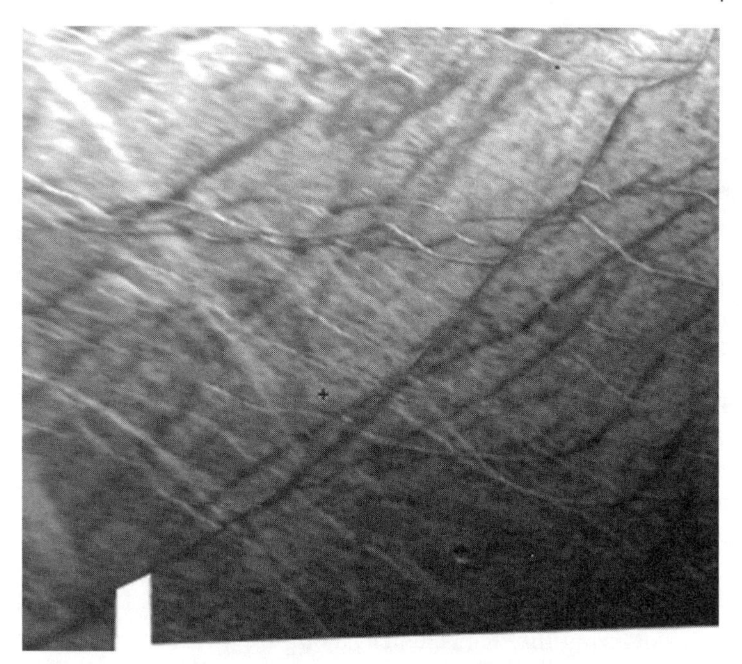

FIGURE 9.1c: Randy Tufts' reconstruction of Astypalaea Linea prior to the strike-slip offset. In the past, as shown, the wispy, old, white lineaments were continuous across Astypalaea, and the angled portions of the fault had not yet been pulled apart by the shear.

images of Europa that had been taken by Voyager in 1979. Amazingly, yet characteristically, the images had been largely ignored for almost two decades after the excitement of Voyager had passed. The discovery made on the basis of these neglected images did not happen until the long-delayed follow-up spacecraft, Galileo, was already in orbit around Jupiter. Ironically, Galileo had a lot to do with the discovery of strike-slip at Astypalaea, though not by capturing the telling images. For one thing, it provided the funds for us to pay Randy's salary as a graduate-student research assistant. And the fact that Galileo was flying around out there stimulated him to look carefully at the older images. But here's the odd thing: Nothing about this discovery really required the Galileo craft to exist. In a similar way, the realization in 1979 that tides could heat the Galilean satellites, which might have come at any time in the twentieth century, was stimulated by the imminent arrival of Voyager at the Jupiter system, not by any data from Voyager itself.

Randy Tufts' discovery had gotten me thinking about the relationship between space missions and science. What lessons do we learn from the fact that some of the biggest discoveries stimulated by space missions were entirely based on earlier data? Perhaps faking nonexistent missions would be a cheap way to get a lot of good research done, but how many times could NASA get away with that? Seriously, the moral is that NASA and other agencies need to invest more in extracting information from the vast amounts of data already on Earth. Too often, valuable space-mission data are archived and forgotten once the short-term excitement wears off. These archived data often contain underappreciated and neglected treasures. Scientists should be encouraged to investigate them. But instead the glory goes to those who jump on to the next new thing.

***O *

Astypalaea Linea lies in the far south of Europa. It branches southward from Libya Linea, passes within about 10° of the south pole, and runs for at least 800 km in all. Voyager images show (**Fig. 9.1a and b**) that, after various widenings and narrowings along its length, at its northern end it widens into a distinct parallelogram shape. It is crossed by a family of ridges, which follow cycloid-shaped trajectories (i.e., chains of connected arcs). Because these ridges cross over Astypalaea, we assume they formed later and so are irrelevant to defining the displacement along the fault. Most significant, as Randy Tufts noticed, are the wispy, old, white lineaments that end at Astypalaea and were evidently cut off by it.

Randy realized that by cutting this picture along Astypalaea and shifting the east side northward along the fault (**Fig. 9.1c**), he could reconstruct the continuity of the wispy lines. What was more, this same displacement closed up the parallelogram. Randy's insightful manipulation showed that the plates along Astypalaea Linea had slid (or scraped) along this boundary for a distance of 40 km. Furthermore, Randy noted that the parallelogram is a classic example of the geometry that must form if the original fault is not straight: The parts of the fault that do not run parallel to the shear direction must pull apart (**Fig. 9.2**). The technical geological term for this feature is a "pull-apart." In effect, the crack dilates along those portions. Randy suggested that the other widenings in Astypalaea, in addition to the most obvious northern parallelogram, might also be pull-aparts.

Pull-aparts associated with strike-slip demonstrate one way that dilational bands can be created. In fact, branching off of the northern end of Astypalaea (to the right, just off the upper right corner of **Fig. 9.1a**) is the dilational band Libya Linea. Evidently, as the huge region to the east of Astypalaea (lower right portion of **Fig. 9.1a**) moved southward (toward the lower left), it pulled Libya open.

***O *

But, then, what drives strike-slip on Europa? Randy Tufts began to address

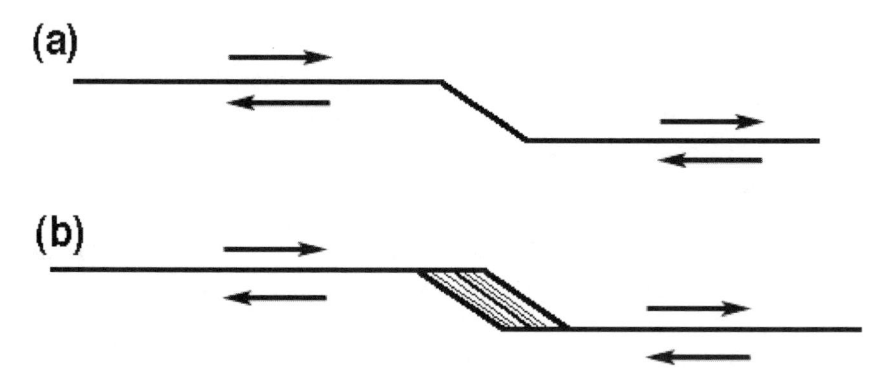

FIGURE 9.2: A schematic (looking downward at the surface) of the general geometric relationship between strike-slip displacement and a pull-apart zone. In part (a), shear is just beginning along the sectors of the fault that run horizontally on the page, but no displacement has yet occurred. Note the jag in the fault between those two sectors. In part (b), shear displacement has occurred along those sectors, but at the jag in the fault, this displacement leads to dilation. In the dilation zone, new material comes up, fills in, and freezes in place. Compare this geometry with the parallelogram in Astypalaea at the upper right in Fig. 9.1a,b.

that problem as soon as he made his discovery of the displacement. He suspected that it might be related to the tidal stresses that my research group had been investigating, but his training in geology was only tangential to the exotic processes that celestial mechanicians ponder.

Fortunately, Randy's background, experience, and talent gave him a much broader and more useful range of expertise than had his formal education alone. In high school in Tucson, he had developed an interest in caves, and a resolution to discover one. This obsession led to undergraduate studies in geology and a systematic program of exploration. His undergraduate years in the late 1960s and early 1970s had other distractions. Randy became a prominent and effective student political activist. But not long after graduation, he fulfilled his high-school goal, discovering with his friend Gary Tenen the large and pristine Arizona cave now known as Kartchner Caverns (but always called by its discoverers simply "The Cave"). For the next two decades, Randy and Gary worked the difficult political problem of obtaining protection for The Cave while keeping it a secret to ensure its preservation. At the same time, Randy pursued a career in social services and public policy. By the late 1980s, the careful political effort paid off and The Cave was on its way to governmental protection.

At that point, Randy began to look for a goal for the next phase of his life of exploration. When he read about Europa and the possibility that there was an ocean there, he realized the implications for possible extraterrestrial life. He decided that he wanted to investigate whether Europa was habitable. Just as he had committed himself to finding a cave, he made a similar commitment to exploration of Europa. And just as his search for the cave included majoring in geology as an undergraduate at the University of Arizona, Tufts enrolled in the PhD program in the same department to begin this new exploration. So one day in the early 1990s, this strange middle-aged man arrived at my office door, and told me he loved Europa. Eventually he became my student and, along with Paul Geissler and Greg Hoppa, part of my core team working on Europa during the Galileo years. Just as Kartchner Caverns was always known around Randy as "The Cave," Astypalaea came to be known as "The Fault."

Randy's political sophistication, honed over all aspects of his multi-faceted career, came in very handy in our dealings with the Galileo imaging team. What's more, his understanding of geology complemented the backgrounds in planetary physics of the rest of my research group. With Randy's geological expertise, we were unconstrained by the canonical positions of the Galileo imaging team's geologists, for better or worse – better in terms of our scientific discoveries, but worse in contributing to our social and political stress. We were encroaching on their turf. Still, Randy's familiarity with the tectonics of deserts in the southwestern US led to his discovery of the Astypalaea strike-slip fault, and no self-respecting geologist could resist the beauty of that discovery.

The discovery had considerable appeal to the public as well, with national headlines like "San Andreas Size Fault Found on a Moon of Jupiter." At about the same time, Kartchner Caverns was opening to great fanfare as a state park. Randy had plenty of use for his considerable media savvy, simultaneously dealing with his terrestrial and extraterrestrial discoveries.

After discovering the strike-slip displacement along The Fault, Randy sat down with Greg Hoppa's charts of tidal stress and meticulously sketched how the stress at Astypalaea would vary over the course of an orbit of Europa. He and Greg shared an office, and they discussed the problem continually for a few weeks.

Fig. 9.3 shows the variation of the tidal stresses at Astypalaea over the course of each orbit, starting (a) when Europa is at the farthest point in its orbit from Jupiter. At this time, tension runs across the fault line. The tidal distortion tends to pull the crack open. One quarter orbit later (b), the stress pattern drives shear along exactly the orientation of Astypalaea. This shear tends to move the region to the lower right of Astypalaea down toward the lower left, the actual direction of displacement observed along

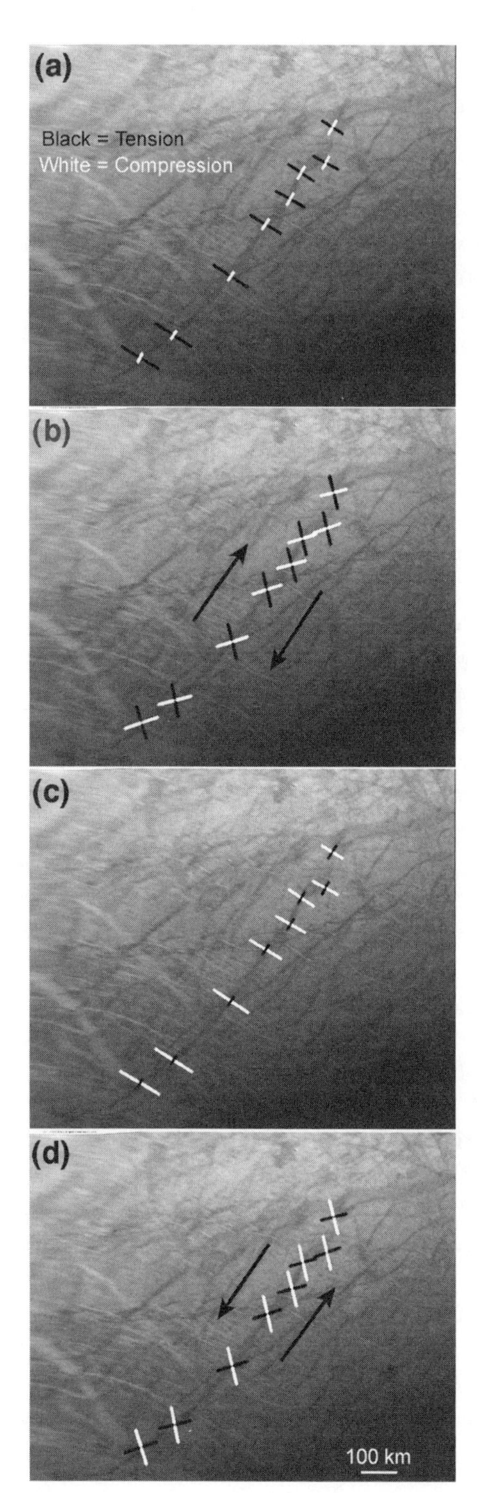

FIGURE 9.3: (a) At one point in Europa's orbit, the tidal stress at Astypalaea is tension across the fault; (b) a quarter-orbit later, diagonal tension and compression yield right-lateral shear (black arrows); (c) a quarter-orbit later the fault is squeezed shut by compression; (d) a quarter orbit after perijove the shear is reversed, but friction in the closed fault prevents slippage. After another quarter orbit, Europa returns to (a). During each orbit, the fault takes a step in right-lateral displacement.

The Fault. In geology jargon, shear in this direction is called "right-lateral"– if you were standing looking across the fault, the terrain on the other side would move to the right. By a quarter orbit later (c), the stress field has reversed itself relative to where we started. Now the sides of Astypalaea are squeezed together by compression. Then, after another quarter orbit (d), the system is in shear, but now in exactly the opposite sense, left lateral. After another quarter orbit, Europa returns to the first point (a).

Evidently, the shear effect is reversed on each cycle, so why wouldn't the strike-slip just move back and forth, going nowhere? The key is that the fault is squeezed shut just before the left-lateral shear, so friction will resist any motion. And the crack is opened slightly before the right-lateral stress, so displacement is free to go in that direction only. The process is similar to walking, where we separate a foot from the ground, move it forward (analogous to shear) just above the ground, then return it to the ground (analogous to compression), then try to shear it back. Friction prevents the foot from moving backward. In exactly the same way, Astypalaea moved ahead in right lateral shear while the crack was held open by tides, and tried to move back in left lateral shear, while the crack was held too tightly compressed to move back. Each day on Europa, Astypalaea took a step forward. Remember, each step was small. Tidal strain each day is only a few meters at most. But with a small step each day, Astypalaea could have moved its full 40-km displacement in only 400 years. The walking would be expected to have continued as long as the crack remained open and active.

The process of ridge-building also seems to have depended on the diurnal tidal working of active cracks, in which crushed ice and slush are squeezed up to the surface. In principle, a fairly large ridge system could form in 1000 years, comparable to the amount of time (400 years) Astypalaea might have walked its full displacement. Those numbers assume perfect efficiency, meaning that all the crushed ice in a crack each day was squeezed up onto the ridge, and that friction never let The Fault slide back at all. In fact neither ridge-building nor tidal walking is likely to be so efficient. But whatever the appropriate correction factors for the two processes, it seems likely that a major strike-slip displacement takes a similar amount of time to building of a major ridge.

Such agreement fits with the idea that both processes depend on how long a crack stays open and active. What would stop the diurnal tidal working of a crack? With sufficient nonsynchronous rotation, any crack is likely to move to a location where the stress pattern is different and can no longer keep the crack from freezing solid. Then both strike-slip displacement and ridge formation could be cut off. It seems that the size of the largest ridges and the distance of the greatest strike-slip displacement are both controlled by the rate of nonsynchronous rotation. Our theoretical models of strike-slip displacement, ridge building, and non-synchronous rotation are at least reasonably consistent with one another.

The tidal-walking model makes a specific prediction for which way any crack should walk, either right-lateral like Astypalaea, left-lateral, or not much at all. The answer depends on the location on Europa and on the crack's orientation (its "strike"). Greg Hoppa and I did the required calculations and found interesting patterns. In a broad belt (60° wide) around the equator, the direction of displacement would depend on the location and orientation of the crack, but further north we predicted only left-lateral displacement, and further south only right-lateral (consistent of course with Astypalaea). We needed to compare this theory with the images.

***O *

Greg Hoppa did a preliminary survey of several locales where images suitable for identifying strike-slip displacement were available. In general, the findings were consistent with predictions of the theory of tidal walking. However, we were soon to discover that some more complicated things must have been going on. In the autumn of 2000, the NASA Space Grant program at the University of Arizona assigned an undergraduate student to do a research project with me. By that time we had an extensive set of Galileo images of Europa, and a complete and systematic survey of strike-slip displacement was long overdue. With undergraduate assistants, you never know what to expect. This time we got Buffy the Vampire Slayer: a tough, smart young woman named Alyssa Sarid, who knew how to work hard, solve problems, and slay monsters as necessary. I assigned Alyssa the task of identifying and mapping strike-slip faults in a systematic way.

Ideally, we would have wanted to survey the entire surface of Europa, but we were faced with a shortage of images. As I mentioned in **Chapter 2,** this shortage was due to the failure of the spacecraft's main communications antenna. In April of 1991, as the spacecraft approached the Jupiter system, it was sent a command to open its larger radio antenna. Until then the antenna had been furled like an umbrella. It was supposed to open into a dish that would allow transmission to Earth of the vast amount of expected data. Each image was 800 $\times$ 800 pixels and the brightness at each pixel was encoded to 256 (or 2^8) levels of gray, so 8 bits of information had to be transmitted for each pixel. The plan was to take about 100,000 images, and the real expectation was that we would take twice as many. Multiplying those numbers, the imaging alone was expected to send about a thousand trillion (10^{12}) bits of data back to Earth. The large antenna was essential, in order to focus enough of the transmitted radio power back to Earth from the great distance of Jupiter, providing the broad communications band needed to handle so much data.

To make a long story short, the umbrella did not open. At least one of the struts must have remained stuck to the center pole. Efforts were made to shake it loose, but nothing worked. The precise cause was never determined, but the mechanism was shop-worn by the delays in the project and perhaps stressed by the three road trips across the US, as described in **Chapter 3**. So the Galileo craft's only working antenna was a small one that had been serving for communications during the trip to Jupiter. At that distance, the tiny antenna could send about 40 bits per second back to Earth—a data-communications rate closer to the standards of a nineteenth-century hand-operated telegraph than to modern broadband equipment. At that rate, transmitting the planned pictures would have taken hundreds of years, well beyond the expected lifespan of the aging camera's image detector, the fuel supply, and the spacecraft itself, to say nothing of the budget allotments and the researchers and scientists themselves. At best, in anything like a reasonable time span, we could get only a few percent of the anticipated images.

The problem was addressed in a remarkable re-engineering project. Galileo's ancient onboard computer was reprogrammed, long-distance, to encode the images more efficiently with modern data-compression techniques. (The computer and its software were so old that the original programmer had to be lured out of retirement to make this scheme work.) Strategies were adopted for averaging over groups of pixels, so that more images could be received, albeit at the expense of resolution. And observing sequences were planned so that pictures were taken quickly during the short fly-bys of each satellite, with data saved to an on-board tape recorder for later slow transmission home during the weeks between satellite-spacecraft encounters.

The canonical evaluation of this effort is that it was completely successful. In a sense it was. The engineering teams at JPL had done a spectacular job of reconfiguring a crippled robot hundreds of millions of miles away and getting the most out of it. The story showed NASA and its teams of engineers working in their best classic form. But ultimately the official spin went too far by suggesting that all of the objectives of the original mission had been met. That conclusion depends on your objectives. As we try to understand Europa, we are stuck with only about 2 percent of the images we had expected. The remarkable self-promotion and euphemizing skills of the Galileo project managers included being able to report 100-percent success with straight faces.

There was some benefit to having the more restricted data set, as Greg

Hoppa often points out. The amount of data was manageable—even one person could go through it in a relatively short period of time. All the data fit on a single CD, and my small research group, consisting mostly of undergraduate and graduate students, was able to finish definitive surveys very quickly. Greg and Randy were familiar with the detailed content of nearly every image of Europa. Such comprehensive knowledge would have been impossible if the data set had been too large. We could never have pulled together a global story so quickly if we'd been faced with challengingly (if not paralyzing) huge numbers of images, like those returned from Mars by recent missions.

That was the upside. The reduction in the number of available images was of course also limiting to the science. Most of the surface of Europa has only been imaged at resolutions of a few kilometers per pixel or worse, as we saw for example in **Color Plates 1–3** and **Fig. 2.3**. Those pictures do not show the details needed for a survey of geological features. We do have high-resolution images, like **Fig. 2.8**, which show great detail, but only at a few selected spots. They do not cover enough of the surface for a survey, and they show places that were selected because they seemed especially interesting to someone for some reason, so they cannot be considered representative. In fact, as discussed in **Chapter 3**, these selection effects can actually make the images even less representative than a random sample, skewing some people's impressions of the character of Europa.

A good compromise between the global coverage at low resolution and the selected sampling at very high resolution is the large set of images that show the surface at about 200 m/pixel. Such images cover somewhat less than 10 percent of Europa's surface. Several of the regions that were imaged at that resolution were selected for their special interest. For example, **Fig. 2.7** is a mosaic of 200-m/pixel images selected because Conamara had seemed so prominent and unusual when it was looked at in global-scale images at worse than a km/pixel. Fortunately, a set of "Regional Mapping" images had been taken at ~200 m/pixel to cover broad, unbiased expanses of the surface for the purpose of geological mapping. For our strike-slip survey we restricted ourselves to these relatively unbiased Regional Mapping images, in particular those covering two broad north-south swaths, each about 250 km wide and running nearly pole to pole. One swath crossed near the middle of the leading hemisphere, the side facing forward in the orbit around Jupiter. The other swath is about 150° further west, on the opposite (trailing) hemisphere.

The completeness of Alyssa Sarid's strike-slip survey was limited in part by the data set and in part by the character of Europa itself. The resolution of the images makes small displacements impossible to see. Moreover, strike-slip faults could only be identified if there were offset features, like the wispy bands that clued Randy to Astypalaea. In many cases, especially for older faults, slicing and dicing has left only short segments, so these faults are less likely to be recognizable, and the survey inevitably favors the more recent portion of the geological record.

***O *

Alyssa's survey revealed dozens of cases of strike-slip displacement. Here I want to point out some details in two of my favorites.

The greatest-displacement champion: In the far north of the trailing hemisphere, a fault at least 200 km long displays a shear displacement of 83 km **(Fig. 9.4a,b)**, nearly twice that of Astypalaea. All faults that far north should be left-lateral according to the tidal-walking theory, and this one conforms to that prediction, just as Astypalaea in the south conforms to the predicted right-lateral displacement. The 83-km offset is the greatest strike-slip displacement so far identified on Europa. This fault is so long and its

FIGURE 9.4(a): One of Europa's largest strike-slip faults, runs over 200 km, diagonally from the upper right to lower left in this unreconstructed image.

offset is so great that, having squeezed it onto the page, we lose much of the detail in the full-resolution version. In addition to its large offset, there has been relatively modest dilation of a few kilometers (Compare **Fig. 9.4a,** which is the original image, with **Fig. 9.4b,** which is the "sliced and shifted" version of **Fig. 9.4a.**) Otherwise, the geometry is fairly simple, because this fault is so straight. Unlike Astypalaea, the original crack was not wiggly enough to produce pull-aparts during the strike slip. While this champion seems shorter than Astypalaea, in fact I have been able to track it in low-resolution over a much greater length than the 200 km seen here. Like Astypalaea, this fault is hundreds of km long.

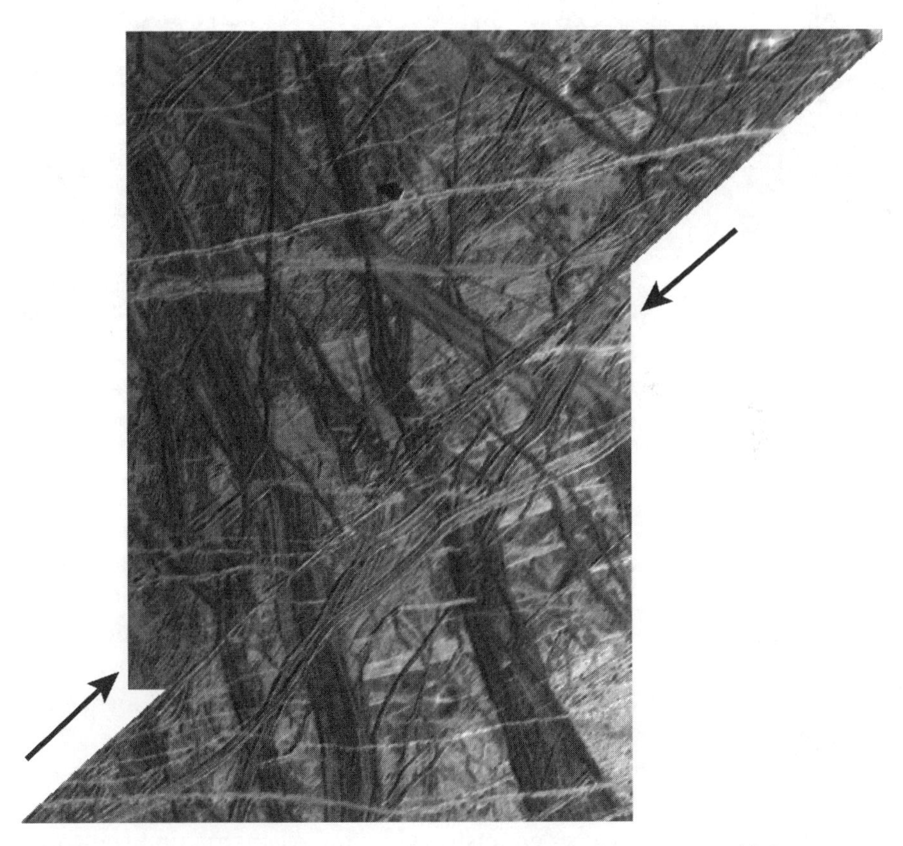

FIGURE 9.4(b): Reconstruction of this fault by manipulation of the image in (a) demonstrates that there has been an offset of about 83 km in the left-lateral sense, just as predicted by our tidal-walking theory. The offset distance is the greatest so far identified on Europa. The reconstruction is based on 200 m/pixel images, which show more detail than can be reproduced in the limited space here.

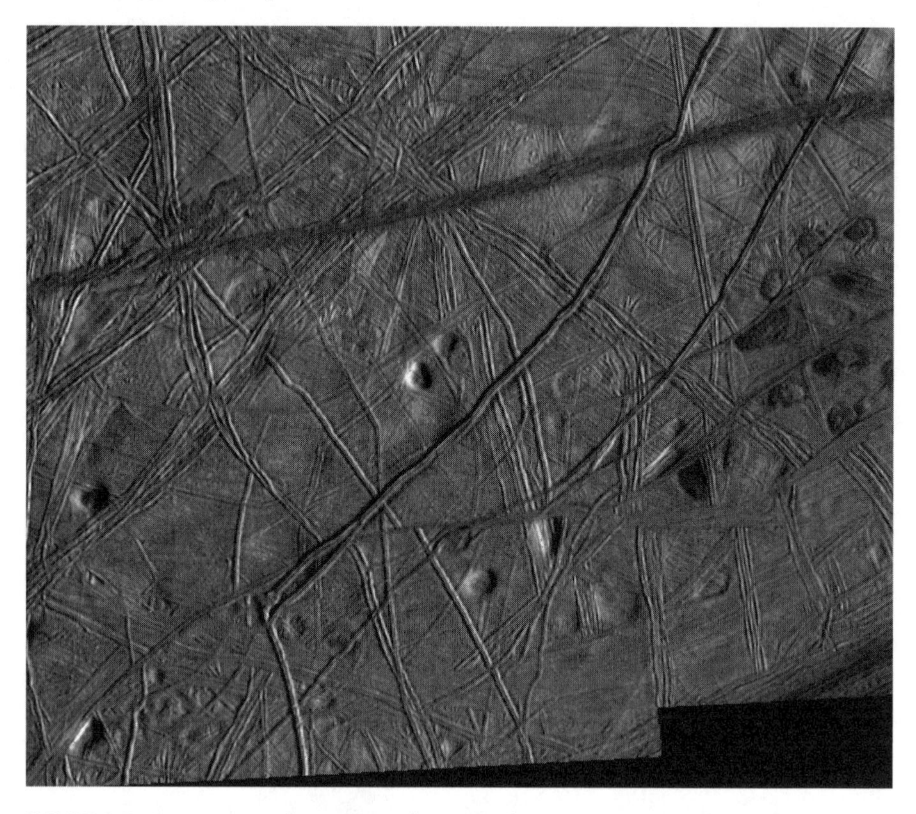

FIGURE 9.5: An area near the equator about 100 km across where there is evidence for a sequence of strike-slip motion along several different fault lines. In Color Plate 4, the evidence for those displacements is marked (in part a) and the history going back in time is reconstructed (parts b, c, and d).

A time sequence of strike-slip: In nearly all the cases of strike-slip displacement that Alyssa mapped, the currently visible fault lines are fairly short and isolated, which is not surprising considering how active this surface has been. Europa's tectonic record shows that the crust of the planet has been sliced and diced severely by various resurfacing processes, including subsequent cracking and ridge building, dilation, and formation of chaotic terrain. Only in rare places can we reconstruct a sequence of interrelated strike-slip displacements. One example is the complex area near the equator on the trailing hemisphere shown in **Fig. 9.5.**

The history of this area is demonstrated by the reconstruction shown in **Color Plate 4. Part a** shows the current appearance of this area, with the strike-slip fault lines in this area indicated in red and labeled with capital letters. Other colored lines show some of the more prominent "piercing" or "cross-cutting" features that help define the slips that have taken place.

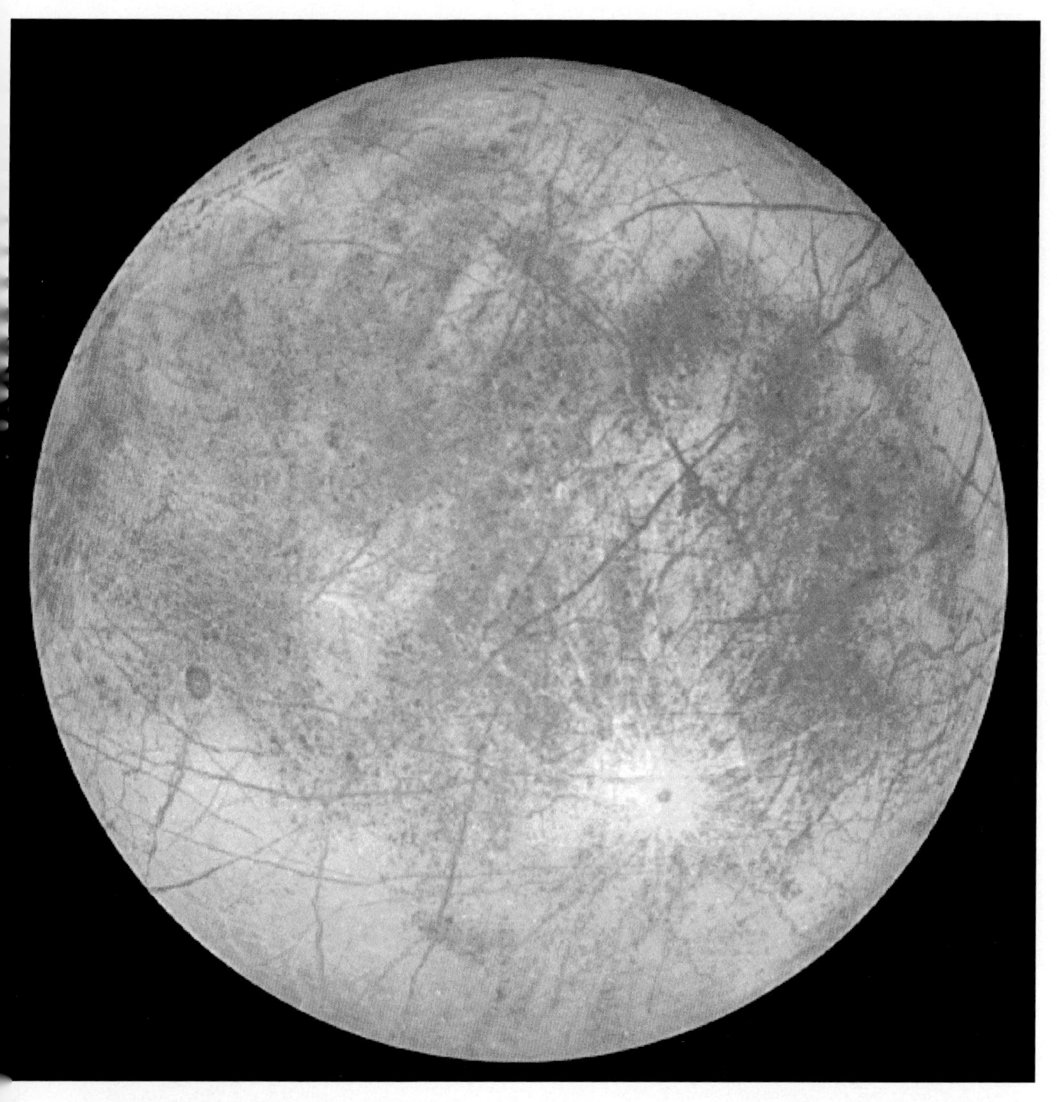

COLOR PLATE 1: A color view of the trailing hemisphere, with the dark (orange-brown) markings enhanced for visibility. The lines mark cracks in the ice crust, while the splotches indicate chaotic terrain. Thus, even at the global scale we see results of the two main resurfacing processes, tectonic stress and heating.

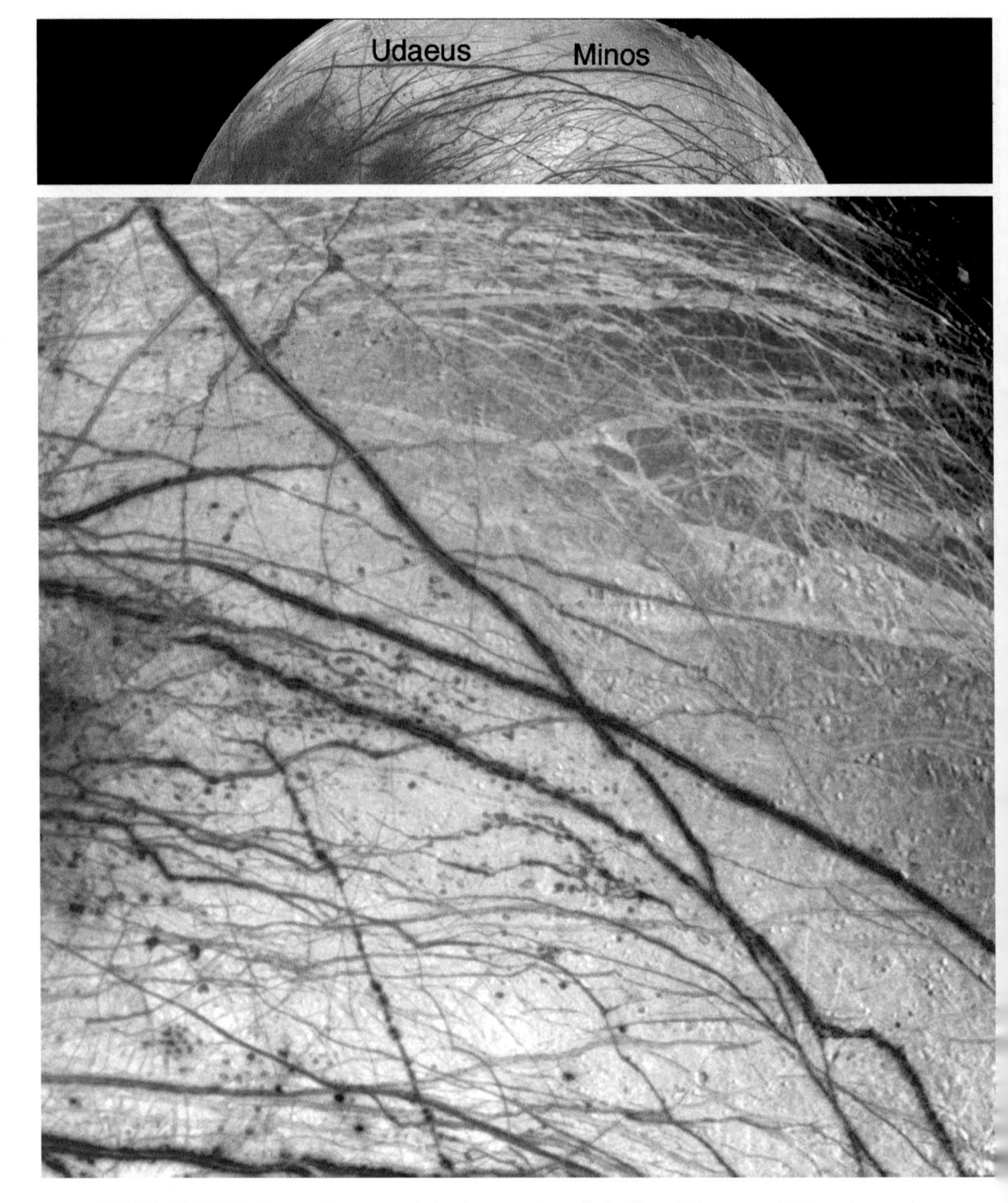

COLOR PLATE 2: The region around the intersection of the lines Udaeus and Minos imaged at 1.6 km/pixel in color. At the top, for comparison, is the same region in a lower resolution global view (part of **Fig. 2.4**). The region also appears edge-on at the upper right of **Color Plate 1**. At the resolution shown here, the global-scale dark lines have bright centers, so they were called *triple bands*. Paul Geissler made this color image using separate black-and-white ones taken through three filters at green, red and near-infrared wavelengths. Paul combined them as if they were blue, green, and red for display to the human eye.

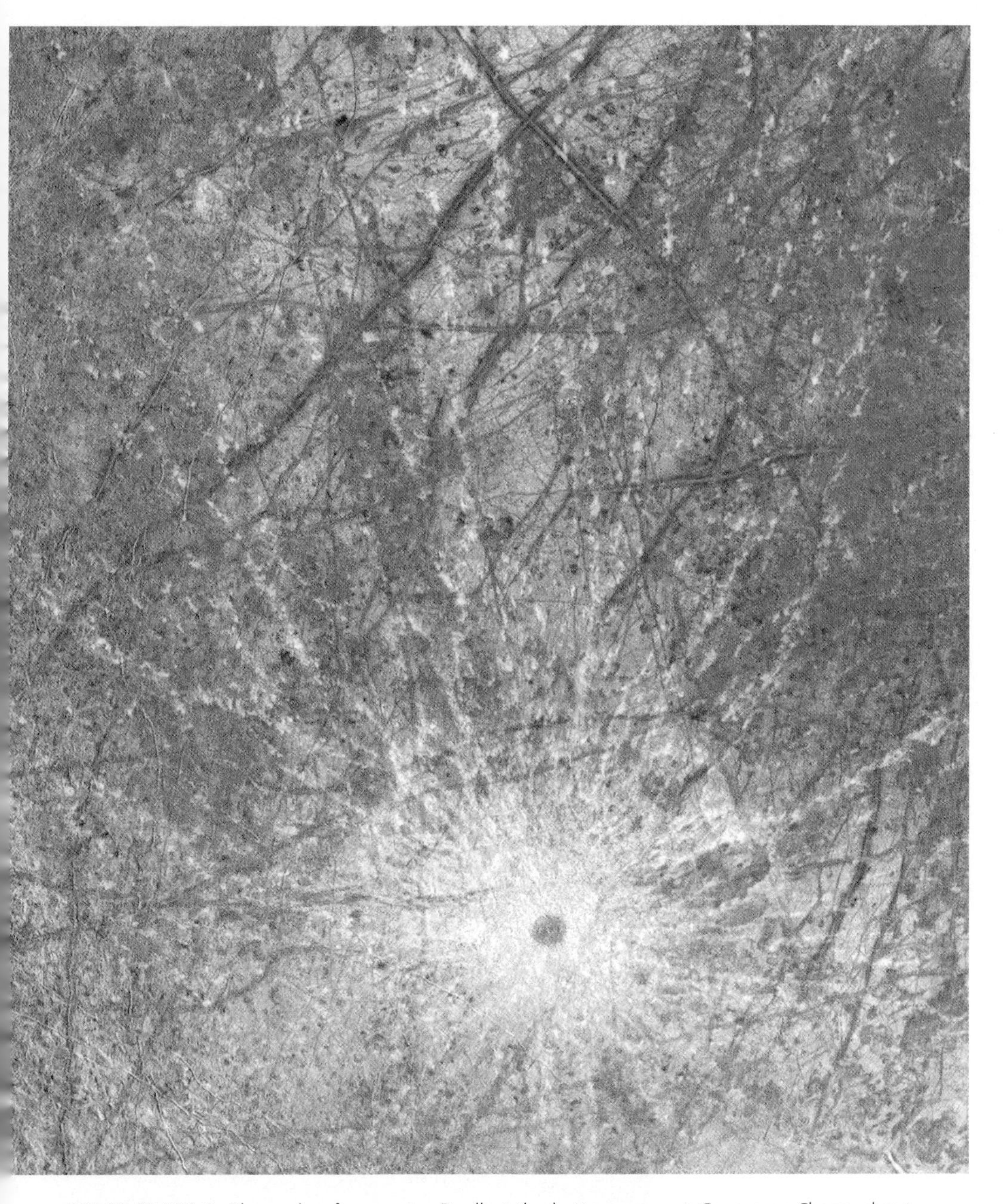

COLOR PLATE 3: The region from crater Pwyll at the bottom up past Conamara Chaos, about 1000 km from bottom to top, imaged during orbit E4. Global scale lineaments are visible with the appearance of ''triple bands''. The ejecta rays from Pwyll extend to, and across, Conamara. Here very-low resolution global color data has been overlaid on the higher resolution E4 images.

(a)

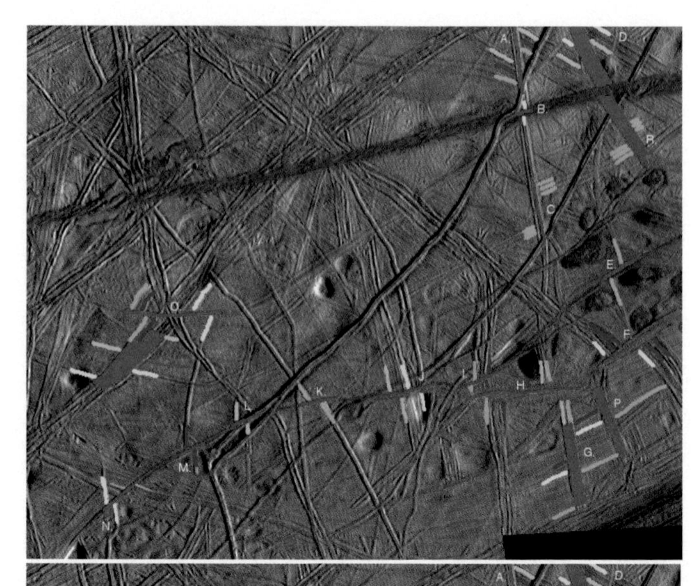

COLOR PLATE 4: A time-sequence reconstruction of a series of strike-slip displacements in the 100-km-wide area shown in **Fig. 9.5**. Part (a) shows the current configuration (as in Fig. 9.5), with faults marked in red and with prominent crossing features marked in other colors (green, yellow, and two shades of blue). The steps go progressively backward in time from the present (a) to the earliest time (d), which is shown on the next page.

(b)

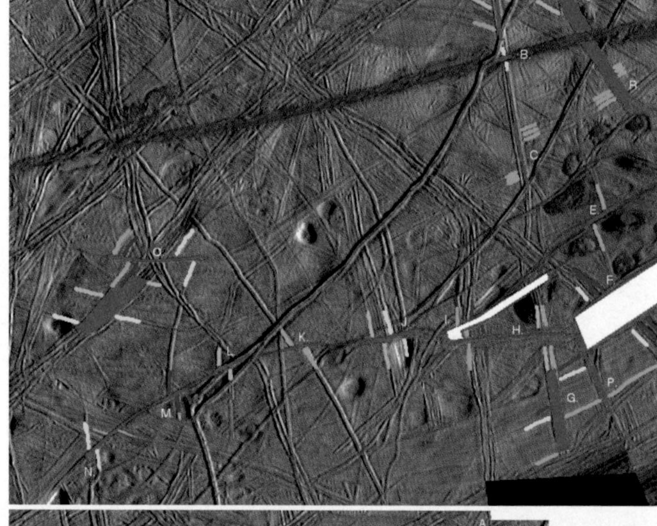

(c)

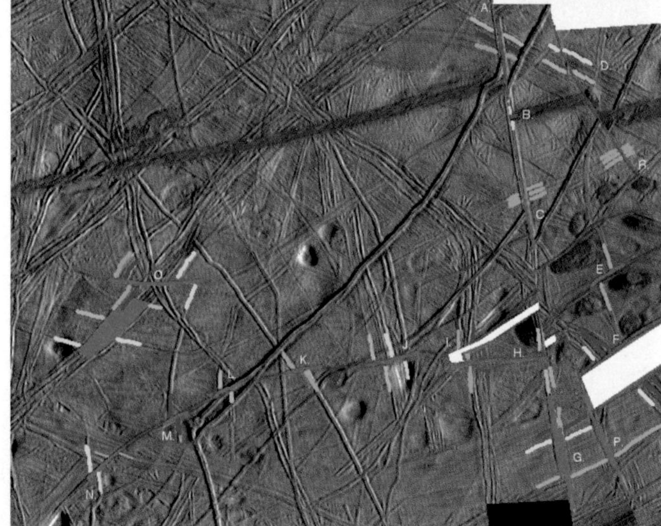

(d)

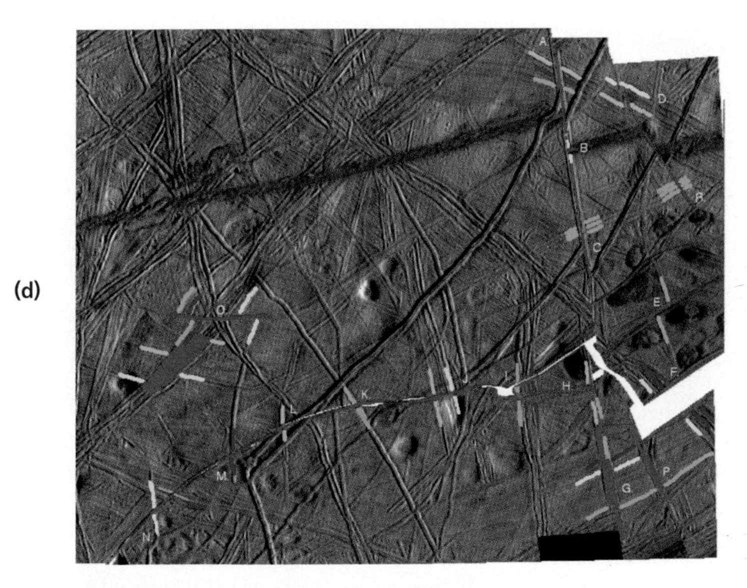

COLOR PLATE 4 (d): A continuation of the sequential steps of reconstruction of the strike-slip displacements, begun in **COLOR PLATE 4 (a-c)** on the previous page. Part (d) is the furthest back in time. It reconstructs the movement along the curved fault JKLMN. As discussed in **Chapter 10**, the white gap that opens at the lower-right as we go back in time represents a place where earlier terrain was destroyed as plates squeezed together.

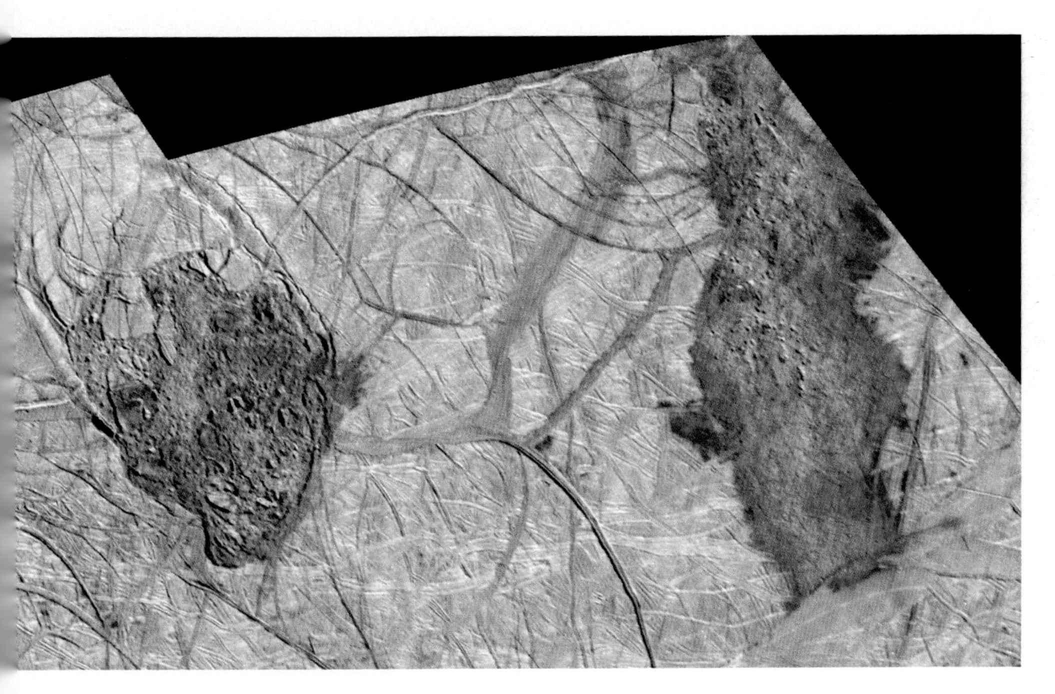

COLOR PLATE 5: Thrace (right) and Thera (left) imaged during orbit E17. For scale, Thera is about 70 km wide. For the location, see the lower right portion of **Fig. 6.1**. These dark patches proved to be typical examples of chaotic terrain, rather than the "cryovolcanic" features anticipated by some geologists.

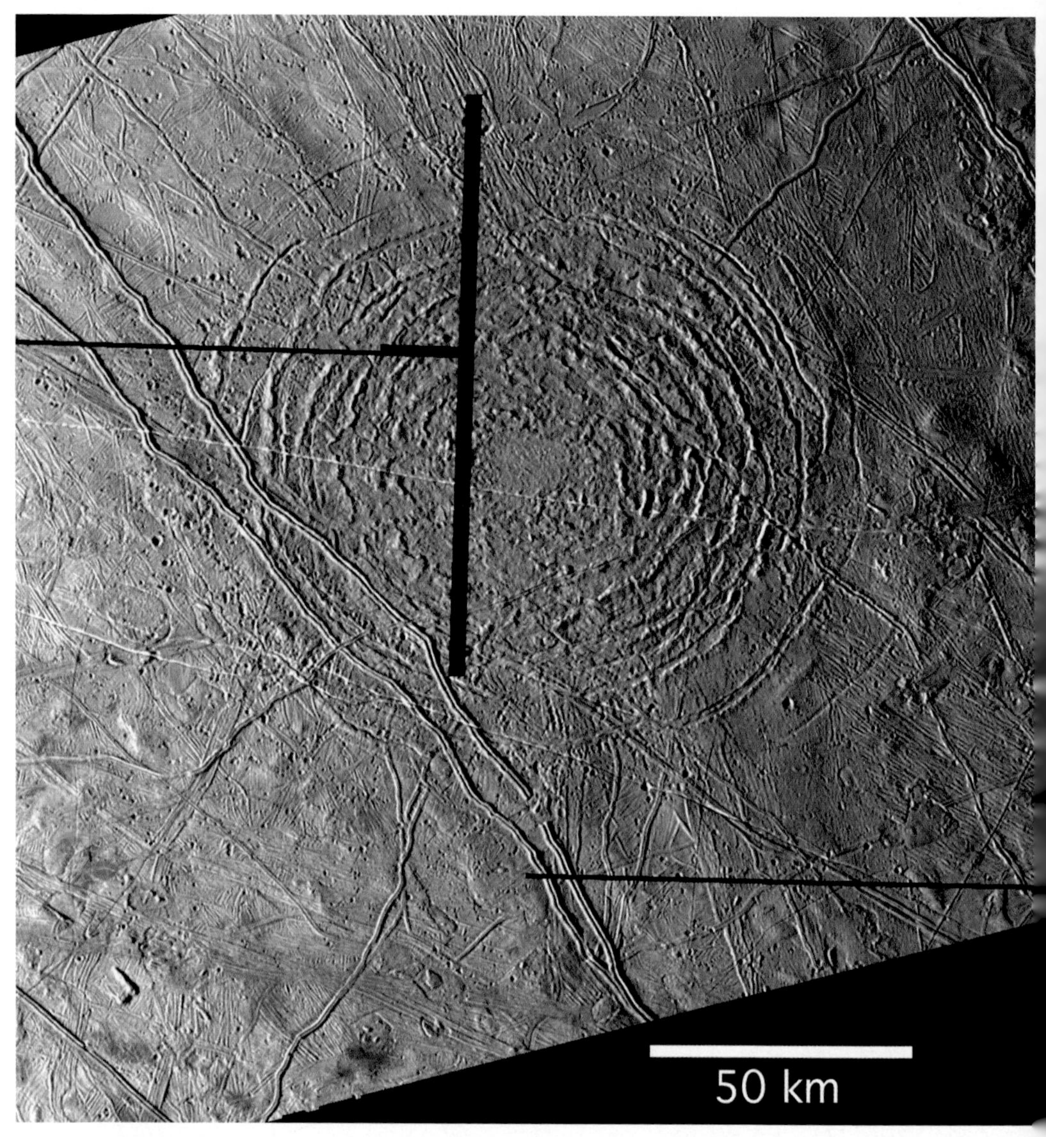

COLOR PLATE 6: A color composite of the impact feature Tyre, from images taken during Galileo orbit E14 at about 200 m/pixel. The inner ring is over 40 km across. The official designation of this feature is Tyre Macula, because it was first recognized and named as a dark area (a macula) on Voyager images, years before its character as an impact scar was revealed by the Galileo spacecraft. Similar to Callanish (**Fig. 15.2**), magnification of this image shows that the surface, where the impactor punctured the ice, is similar to chaotic terrain. (PIRL, University of Arizona).

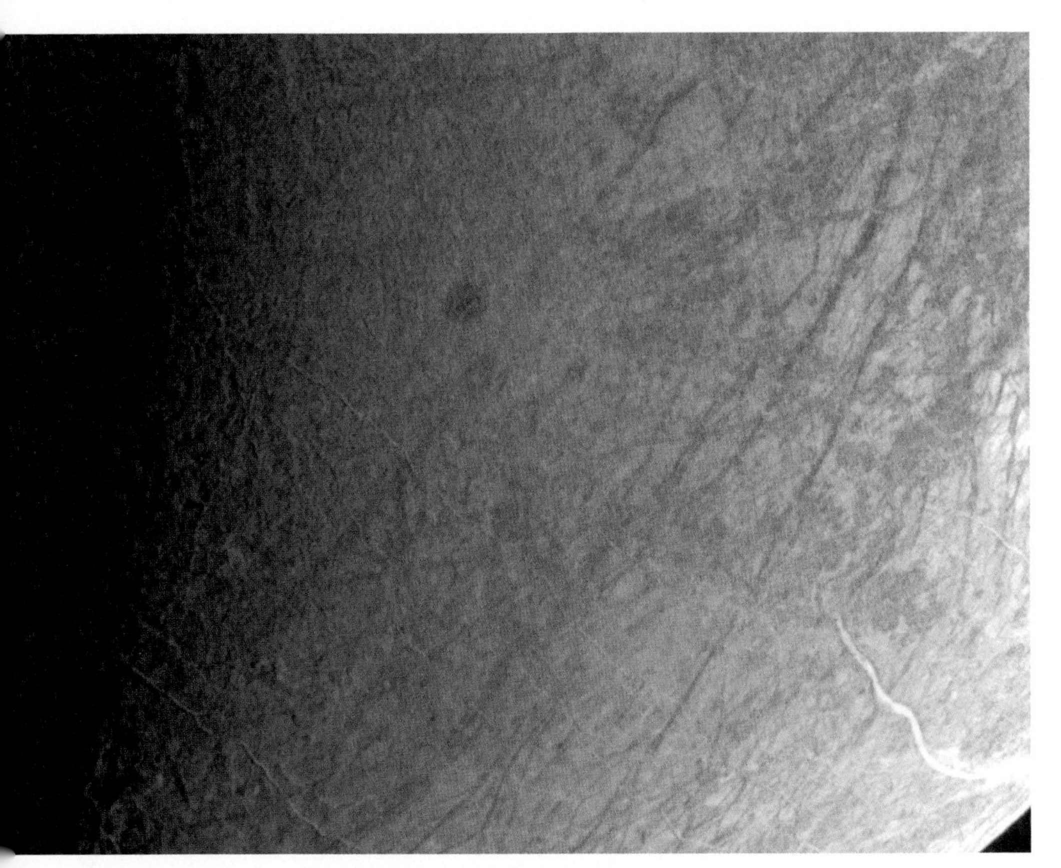

COLOR PLATE 7: Crater Pwyll appears in this broad regional view as a dark orange splotch (to upper left of center), which extends several kilometers beyond the crater itself. The other orange splotches are patches of chaotic terrain. Due to the lighting angle, Pwyll's rays of fine ejecta are not visible here. This color composite by Paul Geissler uses images from orbit E12 taken though infra-red, green, and violet filters and combines them as if they were red, green, and blue. Note the west end of Agenor Linea (**Chapter 10**) at the lower right.

COLOR PLATE 8: One of the most unusual vehicles funded by NASA for space exploration is the DEPTHX submarine robot, shown during testing in the spring of 2007 at Cenote Zacatón, a water-filled, bottomless (until then), limestone cave in Mexico. Its technology for autonomous exploration could be used for navigating open channels into Europa's ocean. This inner-space vehicle indeed bears the NASA logo. And yes, that is duct tape, used to mark the direction of the sampling arm. Photo courtesy DEPTHX Project, Stone Aerospace, and Carnegie-Mellon University. Below, macro-organisms are included for scale (Robin Gary and John Spear photo).

Several strike-slip offset features, labeled J, K, L, M and N lie along a single 100-km-long curved path.

Reconstruction of this area requires several steps because some realignments must be made before others. For example, going backward in time, we needed to realign the strike-slip displacement along faults H and B to straighten out the north-south crack ACG, and only after that was done reconstruct the strike-slip on ACG. The steps of this reverse-time sequence are shown in **Parts b–d** of **Color Plate 4.**

The sequential rearrangement of the surface by strike-slip in this region drives home the sense of continual reprocessing of Europa's surface. Any locale has likely undergone a long sequence of changes, generally with different processes interleaved over time. This particular site provides an especially rich record of a sequence of strike-slip displacements. Moreover it includes a record of other resurfacing processes scattered throughout the period of time represented by the sequence, including cracking, ridge-building, and heat effects—all of which will be discussed in later chapters.

Between **Parts c and d**, we reconstructed the displacement along JKLMN, a 105-km-long arcuate fault. It appears that the adjacent terrain to the south rotated by about 1°, allowing strike-slip to occur uniformly along the entire curve. If it had moved straight, part of the crack would have shown strike slip, while other parts would have dilated. A similar thing happened along DR, where the adjacent terrain rotated just a bit. I cannot explain this phenomenon, but it was fascinating to discover.

Reconstruction along faults P (shown in **Part b**) and G (shown in **Part c**) realigned an older band, one edge of which is marked by a green line. The reconstruction of Fault P also opened up crack F and realigned some ridges (marked by two dark blue lines and a yellow line) that cross crack F.

When I saw that big white gap open up along crack F (**Color Plate 4**), I knew we had the solution to a long-festering problem, introduced in **Chapter 8**: How do we balance the surface-area budget? Some readers may already be able to see why. But let's take a look at another major implication of strike-slip faults before we return to the surface-area budget in the next chapter.

The idea for assigning Alyssa to map strike-slip displacement came from a feeling that something surprising might show up. I wanted Alyssa to compare the actual distribution of left-lateral vs. right-lateral offsets with the predictions of the tidal-walking theory. I had no idea what we might find, but from years of experience, I knew that, if you compare a large new set of data with a solid theory, something interesting is likely to turn up. But to Alyssa, who as an undergraduate was new to scientific research, the purpose of her survey seemed completely vague. After she spent months mapping strike-slip faults, I told her to compare what she found with Greg's predictions. As Alyssa wrote much later, as part of her application to graduate school:

> I then compared the azimuth, location, and sense of strike to the expectations provided by the model. When they did not match up, I first thought that I must have done something wrong. But, I kept thinking about the problem and soon took my first step towards professional research science. I realized that the only way for the data and the theory to coincide was to conclude that Europa's ice shell had undergone polar wander.

That discovery was one of those great moments in a professor's life, when you realize that a student has become a real scientist, or perhaps had been a scientist all along, but was just now showing it. Either way, it was fun guiding her up to that point of discovery.

Here was what Alyssa noticed when I insisted that she look for patterns. In the trailing hemisphere, all strike-slip displacement in the far north was *left*-lateral, exactly as predicted by the theory of tidal walking. Similarly, in the far south, all strike-slip was *right*-lateral, again just as predicted. But then she noticed something that did not fit. The zone where the directions of displacement were mixed, which should have spanned a wide band on both sides of the equator according to the theory, actually ran up from near the equator to about 50°N. The entire swath of surface area in the Regional Mapping images for this hemisphere seemed to be too far north. It needed to be shifted south by about 25° to fit the theory.

One way to reconcile this discrepancy would be to throw out the theory. But the theory was elegant, we liked it, and it explained the general trend (right-lateral in the far south, left in the far north, mixed in between). Naturally, we were reluctant to consign it to the trash.

Since we trusted the theory, it appeared that the terrain in this broad Regional Mapping swath (which extends from the far north to the far south) may have moved roughly 25° northward, relative to the north and south poles of Europa. (The directions of the poles are fixed in space, defined by the spin axis of Europa, just as the position of the poles on Earth are determined by the Earth's daily spin axis.)

In fact it was perfectly plausible that the whole global shell might slip and slide over the ocean in any direction relative to the interior (**Fig. 9.6**). This effect is called "polar wander". If a Santa Claus pole were planted in the ice at the north pole of the spin axis, it would move away from the spin axis if the ice shell slips around. Later, when I needed to explain this geometry to a reporter, I made the sketch shown in **Fig. 9.6** complete with Santa's

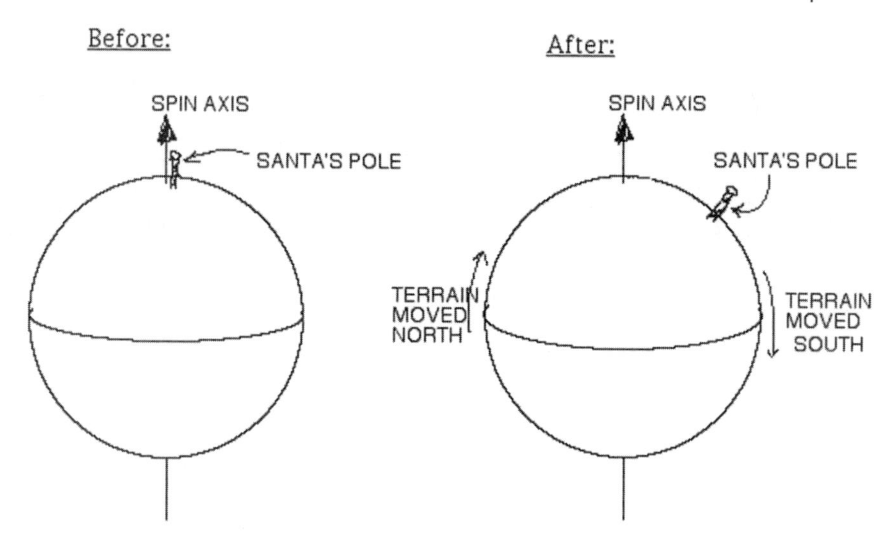

FIGURE 9.6: The ice shell is free to slide around Europa. Suppose Santa's barber pole were planted on the ice at the north pole, right on the spin axis. If the shell slipped as a whole, Santa's pole could move to a new location, away from the spin axis.

FIGURE 9.7: As the ice shell of Europa slips around the body, the poles may move, as they do on Earth when the Arctic ice cap shifts position (Andrew C. Revkin/The New York Times). See Fig. 9.6.

pole, and within a day or two it was all over the web. I was really surprised when a photo of the same thing on Earth appeared on page one of the *New York Times* (**Fig. 9.7**). I knew that the Earth's Arctic ice cap slides around over the Arctic Ocean, but the surprise to me was that there actually is a striped pole up there.

A big difference between Earth and Europa is that the Arctic ice cap covers only the Arctic Ocean, a small portion of our planet's surface, while Europa's ice forms a single continuous unit covering the whole global ocean. And on Europa there are no continents to keep the ice from moving. If the ice were fairly uniform in thickness and density, little force would be needed to reorient the shell in any direction, so it seemed possible that something like that had happened on Europa.

When Alyssa noticed that the long north-south swath in one hemisphere of Europa seemed to have moved northward since all the strike-slip occurred, we guessed that it might have been a case of such polar wander. We reasoned that, if the ice shell of Europa had moved northward on one hemisphere, and the whole shell had moved as a single global unit, then it should have carried the terrain on the other hemisphere southward as in **Fig. 9.6**. Sure enough, Alyssa found that the terrain in that hemisphere has indeed shifted southward, confirming the model of a global slip of the entire ice shell.

We can estimate how recently this shift took place, and hence how rapidly the poles can wander, as follows. The age of the surface is probably less than about 50 million years, but the recognizable strike-slip features are among the more recent tectonic features created during that time. Therefore, the polar wander inferred from those features likely occurred within the past few million years. Moreover, this shift must have happened fairly quickly compared with the period of non-synchronous rotation, and recently enough that not much strike-slip has taken place since. Otherwise the distribution of strike-slip would have again become symmetrical relative to the equator.

What could drive such polar wander? On a nearly spherical spinning body, a very small local excess mass on the surface may cause polar wander, as the site of the anomaly spins outward toward the equator. To make a similar point in a paper in the late 1960s, astrophysicists Peter Goldreich (Caltech) and Alar Toomre (MIT) considered an ant walking on a *perfectly* spherical planet. No matter where it walked, the ant's tiny mass (the anomaly) kept the land under its feet at the equator. (That is, as it moved, the ant kept the equator under its feet by pulling it along; it did not keep its feet on a static equator.) It is unlikely that the mass distribution of Europa's crust is sufficiently symmetric that a bug's mass would spin out to the equator and drag the icy shell along with it. However, the ice crust is very thin and has fairly low topography and is probably detached from the interior by the underlying global ocean. (The polar wander theory requires that the ice crust be effectively uncoupled from the solid interior, so the crust can slip as a single unit. Substantial grounding of the ice layer, for example on invisible rocky continents sticking up from the sea floor, would be ruled out.) An unusually thick region or a site of a high bump on the surface might be pulled by centrifugal force out toward the equator, dragging the entire ice shell with it. And in fact one of the biggest bumps on the surface (described in detail in **Chapter 14**) does lie on the equator. Moreover, theoretical models of tidal heating could cause regional thickening of the ice that could affect polar wandering. We cannot be certain, but any of these mechanisms could play a role.

***O *

Study of strike-slip faults has taught us a great deal about Europa. It demonstrates a direct link between diurnal tidal distortion of the satellite and tectonics through the process of tidal walking. It provided evidence for polar wander. It even provided a key clue to balancing the surface-area budget, an important issue to be discussed in the next chapter.

The tidal walking mechanism places constraints on the structure of the crust. First, the model that we developed needs to slide over a low-friction layer, the ocean. It also requires that the cracks must penetrate to that layer in order to allow the daily steps of tidal walking. Penetration of cracks means that the ice must be quite thin, probably less than 10 km, because tides are probably too weak to drive cracks through thicker ice. Greg Hoppa likes to point out, too, that strike-slip is only found along cracks that have developed ridges, not along cracks without them. That observation is consistent with our models of ridge formation and of strike-slip walking. Both require cracks that penetrate to liquid.

Also, our models for tidal walking and ridge-building both require that the cracks stay active for about the same amount of time, at least several thousand years. That result supports the idea that they both depend on the same thing, actively working cracks that penetrate to the ocean. It also means they are both shut down by the same thing, probably non-synchronous rotation to where the tides are different. These considerations suggest a rotation period of about 50,000 years (give or take a factor of a few) relative to the direction of Jupiter. (But, relative to a fixed direction, such as toward a distant star, Europa's rotation period is only a few days, practically the same as its orbital period.)

People who are vested in the thick-ice party line have a hard time dealing with the idea that cracks penetrate through to the ocean. Their assumption has been that the processes we see recorded on Europa's surface involved solid ice, independent of the liquid ocean below. But that would require some process as effective as our model at driving strike-slip, and as successful at fitting what we observe on the surface.

Recently, I asked a graduate student, Brandon Preblich, to try doing some computer simulations that might test whether tidal-walking could occur in thick ice. The idea is that the warmer, more viscous ice deep under the surface might flow enough to replace liquid water as the lubricant to allow the surface to move. Brandon was able to get some tidal walking to occur, but only with very contrived conditions. In general, tidal walking needs the crack to penetrate down to the ocean.

The crust of Europa seems most likely to be very thin ice over liquid water. As studies of the tectonic record continue, in concert with better theories of the underlying processes, we could learn much more about the history and structure of Europa. So far only the most recent part of the tectonic record has been exploited, but future research should be able to penetrate even further back in time to reveal more about the complex dynamics of Europa's crust.

The party line is that we cannot learn much more about Europa until we send another mission. Specifically, it is claimed that only with future spacecraft can we address whether the crust is so thick that the ocean is isolated from the surface or is thin enough that the ocean is intimately linked to the surface. That line is designed to sell space missions. The reality is that shortly after each major planetary mission, the data are set aside and ignored. We have already seen numerous examples of how old spacecraft data provide major discoveries decades after they were received, when someone bothers to go back and take a look. Randy Tufts' discovery of the strike-slip fault Astypalaea in old Voyager images is just one example. Instead of mothballing data shortly after the mission press conferences, we should be pursuing the various lines of scientific evidence. We can continue to make major discoveries about Europa, even without waiting for the next mission. And the mounting evidence indicates that Europa's icy crust is permeable, with multiple connections from the surface to the liquid ocean.

Convergence 

Europa's crust is highly mobile. Tides created the cracks that allow plates of crust to slide around, and tides formed the ridges that mark those cracks. Thanks to those markers, and to the fact that there are no trees or other annoying manifestations of life to get in the way of our views, we can reconstruct the past motions of those plates.

Even the limited Voyager images showed clear evidence of both strike-slip displacement and dilation along cracks. Although Schenk and McKinnon had encountered resistance to the seemingly radical notion that a surface could be so mobile, Galileo images confirmed that dilation had been common. Large crustal plates (often hundreds of kilometers in scale) have separated, with the space between them filled by new surface material, creating the dilational bands discussed in **Chapter 8**. Similarly, new surface area has been created along strike-slip faults, where bends in the fault lines have resulted in pull-apart zones (**Chapter 9**). Whether dilation is caused by separating plates or by slip-strike pull-aparts, it increases the surface area of Europa's ice crust as new material comes up to fill the gaps.

In contrast, most other resurfacing processes on Europa do not add (nor for that matter subtract) surface area. In terms of what I like to call the surface-area "budget," they are revenue-neutral. Consider ridge formation, in which tides pump slush out from a crack over the adjacent terrain. New surface is created for up to a kilometer on either side of the crack, often covering hundreds of square kilometers along the full length. But every bit of new area in these ridges covers up an exactly equal amount of older surface. Likewise with chaos formation: Previous terrain is disrupted and melted and the new surface (matrix and displaced rafts) forms entirely within the existing area of the destroyed terrain. There, too, precisely as much old surface area is destroyed as new surface is created. Dilation, though, is a completely different story: As a crack opens, new surface area is created, but in the process there is no obvious covering up or removal of existing surface.

Because new surface area has been continually created on Europa in this way, someplace and somehow crust must be removed to compensate for it.

In some way, the surface-area budget must be balanced as new surface is created at dilation zones. A nagging question for several years was, how and where has crust been removed so that Europa's total surface area is conserved?

On Earth, we have the same issue, with rapid dilation occurring on a global scale, including the opening of the Atlantic Ocean, the separation along the East African Rift, and the parting of the Red Sea. With so much dilation between crustal plates creating new area, several processes help keep Earth's surface-area budget balanced. Most common in large-scale plate tectonics is plate subduction, where one plate of the Earth's crust rides down under another. The topography that marks the locations of subduction is very dramatic and distinct, although usually hidden by an ocean. Between a subducting plate and the edge of the plate that overrides it, the sea floor drops into deep narrow trenches, the deepest spots in the ocean. No such trenches have been identified in the icy crust of Europa.

When plates of continental crust, as opposed to oceanic crust, converge on Earth, they are too thick for one to slide under the other, so the material must pile up, notably when India crashed into southern Asia, creating the Himalayas, but no such mountain ranges have been seen on Europa.

Surface area can also be taken up by thrust faulting, where a dipped fault plane allows one plate to ride slightly up over another, but not nearly as much as in the case of subduction. Thrust faults are a common way that horizontal compression is relieved on Earth, and they are common and large enough on Mercury that they may be part of a global contraction process. But again, there is no evidence for thrust faulting on Europa.

Europa's surface-area budget has an enormous income of newly generated surface, but where does it all get spent? The reconstruction of strike-slip displacement shown in **Color Plate 4** gave a first hint regarding what I suspect is the dominant "sink" for excess surface area. I return to that story in this chapter. But first I want to describe a couple of earlier proposals for how excess surface might be removed.

The horizontal compression created by the surface spreading at some locations could be accommodated elsewhere by corrugation of the surface in a series of crests and troughs. Imagine a kind of accordion-fold crumpling, which contracts surface area. In principle, such corrugation could help balance the surface-area budget, reducing area while, at dilation sites, new area is created. Such corrugations, with a spacing of about 25 km, have been observed in one site, which lies within the pull-apart zone of Astypalaea. Fine cracks there may correlate roughly with crests between the troughs. The fine lines might have resulted from the bending of the crust over the crests, although the correlation between those cracks and the crests of the corrugations is really not very accurate. The evidence for corrugations on Europa consists of just these three waves in this single, rather special location.

Louise Prockter and Bob Pappalardo, from Jim Head's Brown University group, assumed that these corrugations were created by horizontal compression of the crust. They reported that tidal compression could not have created corrugations of 25 km wavelength unless the surface was riding over a thick layer of viscous ice that was distorted below it. They concluded that they had found the mechanism for accommodating horizontal compression.

Prockter and Pappalardo's theory of folding was based on their belief in thick ice, so the discovery of the folds was promoted as evidence for the canonical model of an isolated ocean. In their word, the result "discounts" the view of Europa that had been emerging from my research group's work on tidal tectonics in which the ice is thin enough that the ocean is linked to the surface. Their entire paper on the subject was only two pages long, and the proposed theoretical model was described in four sentences. Such brevity can be justified by the format of *Science* magazine, where it was published in the summer of 2000. *Science* readers usually assume that a transparent presentation of the details will be published elsewhere, in a more specialized journal. But in this case no details were forthcoming. Nonetheless, in a familiar pattern, the story was fast-tracked into the canon without any serious consideration of potential shortcomings.

The story is interesting, but I can see why no details were ever published. Consider the observations. Three wavelengths in a special locale (entirely within Astypalaea's pull-apart zone) do not a globally important phenomenon make. The features *are* interesting and provocative, but ordinarily such oddities would not be considered the basis for such far-reaching implications. Prockter and Pappalardo pointed out two other examples of cracks or troughs, perhaps similar those seen in Astypalaea, but examples with less than one wavelength are hardly convincing signs of corrugation. The observational evidence supporting their model is marginal.

Now consider the theoretical part of their story, at least to the extent we can, given that hardly anything about it is published. The key to its purported success is that the amount of stress required could supposedly be provided by tides. Prockter and Pappalardo needed to invoke several tens of degrees of non-synchronous rotation to accumulate the required stress. By our estimates, that much rotation would take at least thousands of years. Remember, ice is like silly putty. It acts elastic and brittle if you distort it quickly, but soft and viscous if the change is slow. That is why glaciers can flow. Europa's crust would flow during the slow nonsynchronous rotation, so the stress imagined by Prockter and Pappalardo could never build up.

Finally, the effectiveness of corrugations at reducing surface area suffers from a basic problem of geometry: The shortening of the surface area is negligible unless the slopes of the corrugations are fairly steep. The troughs that Prockter and Pappalardo noticed in Astypalaea slope up toward the crests on a grade of about 1 percent. You would hardly notice it if you were walking up the hill. The shortening of the surface would be less than 1 percent of 1 percent. Even if such corrugations extended for 1000 km, they would account for negligible surface contraction.

If corrugation plays a significant role in the surface-area budget, a case for it has yet to be made.

***O *

By the time Prockter and Pappalardo's *Science* paper came out, the Galileo political situation had already crystallized. So I was not surprised that, despite its lack of credible underpinnings, the story was immediately accepted into the party line. It was the word of the designated Galileo Europa spokesman, and it claimed to demonstrate that the ice was very thick. So, according to the party line, the problem of how to compensate for the formation of new surface was solved by the corrugation model.

For all the scientific reasons I described above, I wasn't buying it. Besides, I already had my own favorite place to hide extra surface area. Nearly half of Europa's crust is chaotic terrain, and most of that terrain is in very small patches. The crust has been continually perforated by these openings, which could readily accommodate considerable amounts of surface contraction. Consider the analogy of a sheet of heavily perforated metal; its area could be contracted leaving minimal evidence of distortion in the metal, because the holes would take up the distortion. Each hole would change size and shape a little bit. For example, a round hole would squish into an oval shape. The distortion of the metal itself would be minimal, with little if any visible effect. In the same way, holes in Europa's crust, while open during the formation of chaotic terrain, could have accommodated considerable surface contraction without creating obvious signs of distortion.

In this lateral compression process, each opening in the ice would have its size reduced and its shape distorted slightly. But, because each already had an irregular shape, when we look at the chaotic terrain that now fills in each hole, we have no way to recognize whether the preceding opening changed in size or shape. So, even if the openings had accommodated the lateral compression of the surface, it would have happened without leaving a recognizable trace.

Several of the people assigned to denigrate the thin, permeable-ice model have mocked this explanation. Their line has been that if we cannot point to visible effects of the horizontal compression, this mechanism has no merit. I find that logic baffling. We know that a huge fraction of the area (nearly half) of the crust has been perforated through by chaos, and we know that such openings can relieve stress and allow an area to contract. Chaos would provide a perfectly plausible explanation for how surface area is reduced in one area to compensate for dilation elsewhere.

So much of the surface has been continually perforated by chaos that it would be surprising if these openings did not play a big role in accommodating the spreading of dilation bands. Whatever process has been operating over broad regional scales to pull apart dilation bands (as discussed in **Chapter 8**) has probably also been squeezing chaos openings at the same time. Future study of the geological record might show whether dilational bands tended to form at times when chaos openings happened to be particularly common. Eventually, we may be able to determine how much of a role chaos had in allowing regions of the surface to contract under horizontal compression.

***O *

The strike-slip reconstructions in **Color Plate 4** produced a big white gap in the picture as we went back in time. Think about that result: One linear band (F) was about 8 km wider in the past than it is now. It dilated *going back in time*. So, going forward in time, this must have been a site of the opposite of dilation. This is a place where plates of the crust were pushed together, rather than pulled apart along cracks. Where plates are pushed together on the Earth, they usually pass above and below one another (causing earthquakes and volcanoes) or, where continents collide, there are giant collisions (most recently building the Himalayas). Here something quite different has happened.

Once we discovered this site where the opposite of dilation occurred, we were not sure what to call it. Geologists did not seem to have a generic term for such negative dilation. I decided to use the word *convergence* to represent displacement where plates of crust have moved together as surface area at their shared boundary has disappeared. I try to avoid the word *compression* here, because that would imply something about force. Here we are simply describing the geometry of the displacement. It is, to say it again, a place where surface area has disappeared. Convergence was inferred to have occurred at this site because it was necessary to accommodate the observed strike-slip displacement.

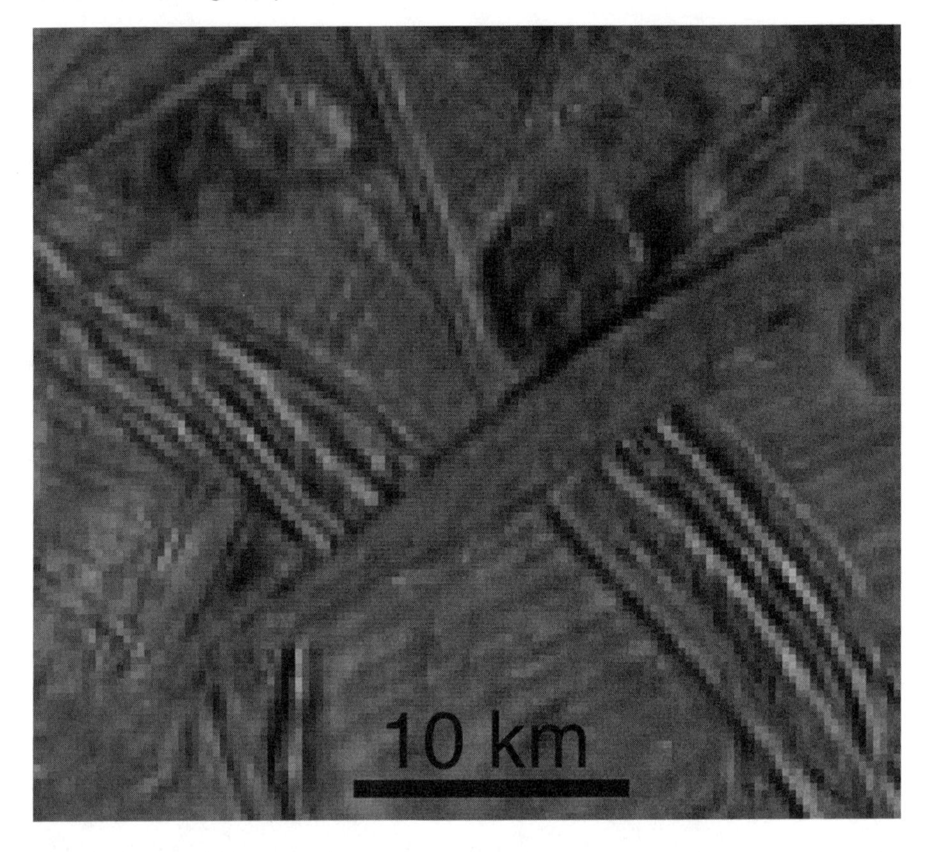

FIGURE 10.1: The "Himalayas of Europa" appear in this enlargement of the convergence site at location F in Color Plate 4. This feature is visible (but shown smaller) in Color Plate 4(a), while the reconstruction produces a wide white gap in the past (Color Plate 4(d)), which shows that convergence has occurred here.

This site provides us with an example of what kind of mark convergence leaves on Europa (**Fig. 10.1**). In appearance, it is band-like, but unlike dilational bands the opposite sides do not fit together. In addition, one side (to the north) has a slightly raised lip, which may have plowed into and over some of the adjoining terrain. This site looks like a place where plates have scrunched together in a collision. These "Himalayas of Europa" are a few kilometers wide and 20 meters tall.

But wait, there's more. Another strike-slip reconstruction revealed a site of convergence in another place. **Fig. 10.2a** shows the equatorial part of the Regional Mapping images in the trailing hemisphere. A 600-km-long, bent, cycloid-shaped fault runs across the area, as shown in **Fig. 10.2b**, with dilated sections and misaligned crossing features. We can perfectly reconstruct this fault by rotating the huge area on the east side by about 1° (**Fig. 10.2c**). When

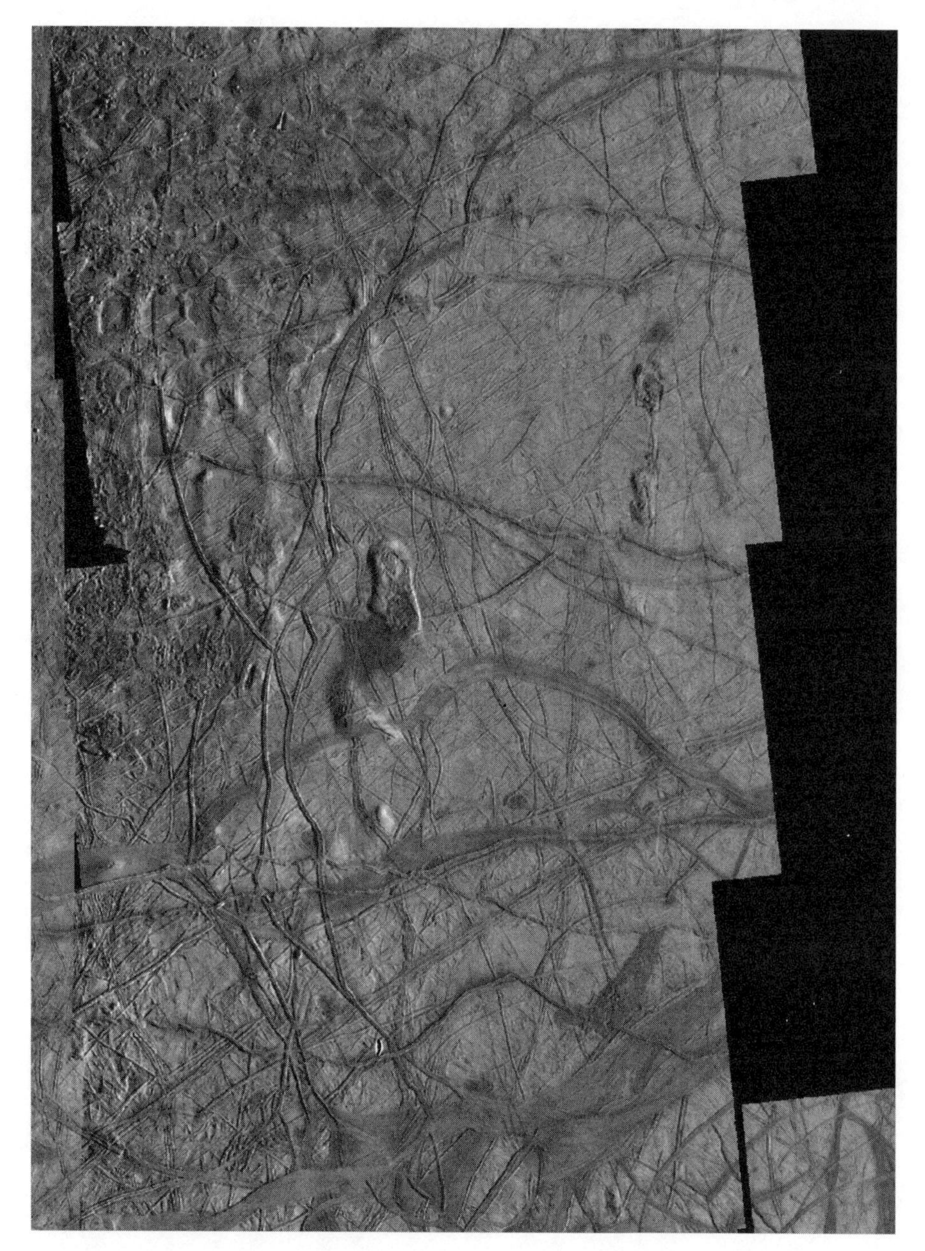

FIGURE 10.2(a): The equatorial portion of a mosaic of E15 Regional Mapping images.

this plate is rotated, it pulls away from the terrain further south, opening an 8-km-wide gap, well over 100 km long. Going forward in time, this geometry represents another convergence zone.

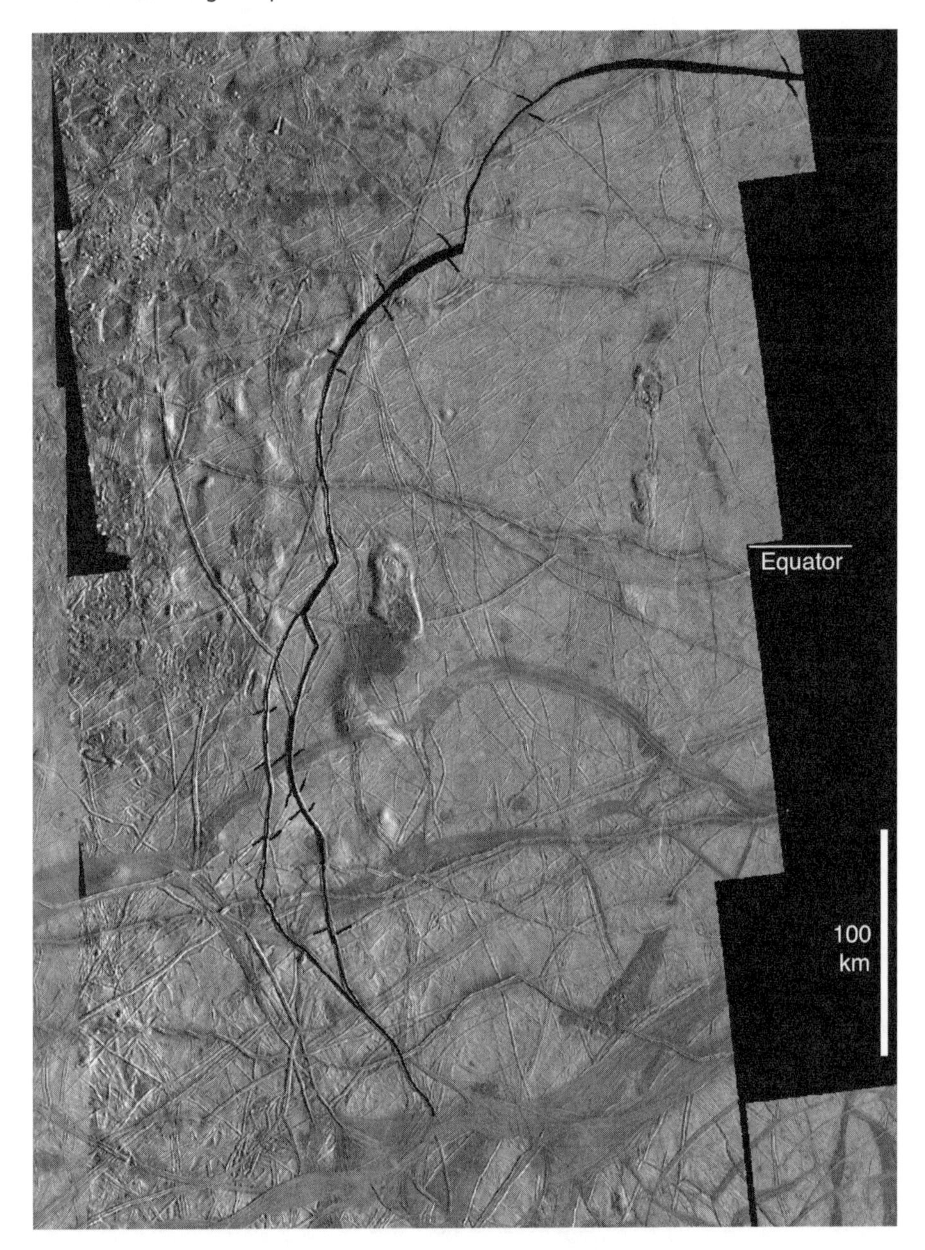

FIGURE 10.2(b): A large, bent, cycloidal lineament is marked, showing several prominent indicators of strike-slip displacement.

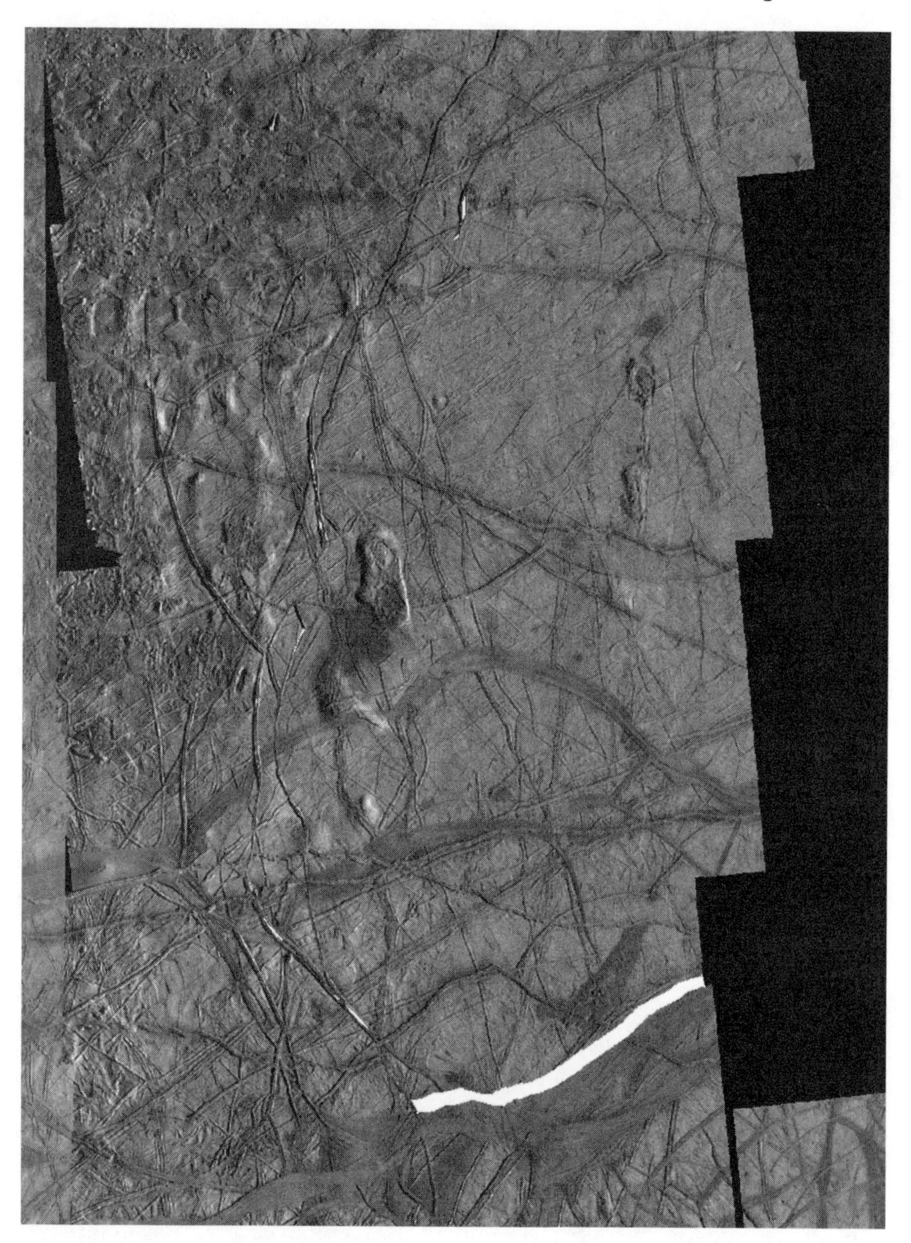

FIGURE 10.2(c): Reconstruction (involving a 1° rotation of a huge plate of crust to the east) shows excellent fit along the fault. It creates an 8-km-wide gap at the south, where surface convergence has occurred.

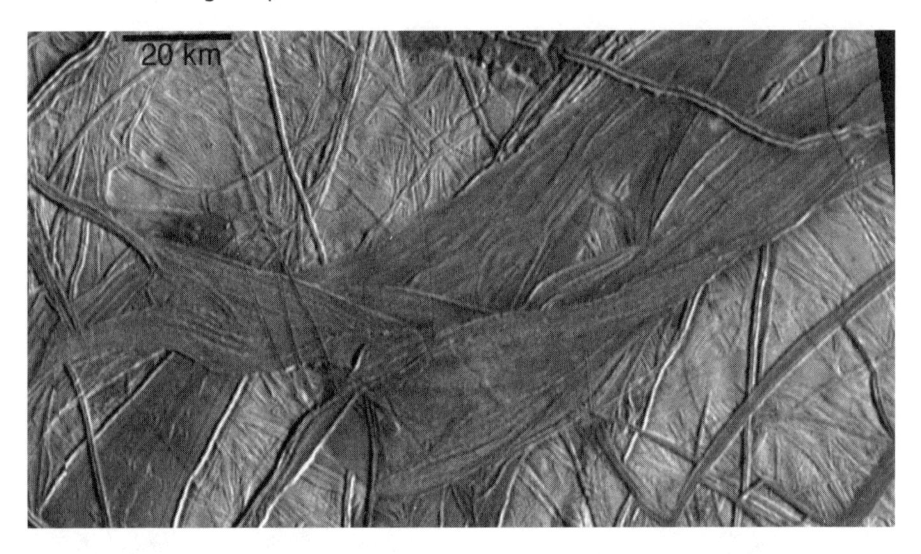

FIGURE 10.2(d): The area of convergence, inferred from the reconstruction in (c). Note the similarity to the convergence site in **Fig. 10.1**.

The convergence sites in **Figs. 10.1** and **10.2d** look very similar to one another. The convergence bands consist of segments with fine internal striations, like dilational bands, but whose sides are not parallel or reconstructable. These shapes remind me of microscopic views of animal muscle tissue. Each segment appears wider in the middle, and often has a slightly raised lip on one side. Even where we are missing the fortuitous clues needed for reconstruction, we now have a way to recognize likely sites of convergence, based on this characteristic appearance.

***O *

One linear feature on Europa with similar characteristics had already been nominated as a convergence feature on the basis of independent, circumstantial evidence. Agenor Linea is nearly 1400 km long, running roughly east-west, south of the Wedges region (**Fig. 10.3**). The similar, smaller Katreus Linea lies parallel to Agenor and likely formed the same way.

Three characteristics suggest that Agenor may be a convergence feature:

- First, Agenor has resisted efforts at reconstruction. For dilational bands reconstruction is often obvious because neighboring plates fit together like pieces of a picture puzzle. On the other hand, by definition, at a convergence site the adjacent terrain from neighboring plates is missing. Hence, the mismatch of the crust on opposite sides of Agenor could be explained by convergence.
- Second, the proximity of the Wedges region (**Fig. 10.3**), where dilation

FIGURE 10.3: Indicated by the lower pair of facing arrows, Agenor Linea is the bright band running across the lower portion of this region. Katreus Linea, bracketed by the upper pair of arrows, is the smaller, similar bright band just north of Agenor. This image is enlarged from the Mercator-projected mosaic by USGS (**Fig. 6.1**). It extends from about 7°N (at the top) to about 45°S, and from longitude 170° to 250°W. Most of the northern half of this image contains the Wedges region, which lies between the mottled-appearing chaotic terrain shown to the left and right.

has been a source of plenty of new surface, is consistent with Agenor being a site of corresponding surface-removal by convergence.

- Third, Agenor's appearance at higher resolution (**Fig. 10.4**) is similar to the high-resolution views of convergence zones we found from the strike-slip reconstructions, with their curved boundaries and the non-parallel, "muscle-tissue" striations.

While several pieces of evidence collectively point to Agenor (along with the associated Katreus) being a convergence feature, they do not prove it. For further evidence, however, we have to seek out the Evil Twin of Agenor.

***O *

I spotted the Evil Twin on the Jupiter-facing hemisphere of Europa , which has generally been imaged at low resolution. The best we have is a mosaic of a dozen images taken at about 1 km/pixel during Galileo's twenty-fifth orbit.

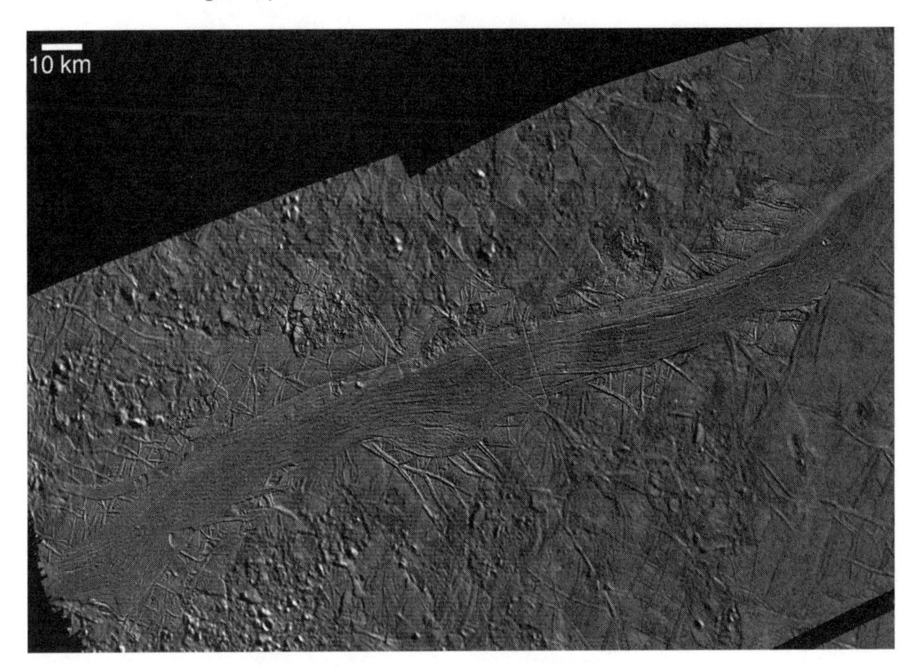

FIGURE 10.4: This high-resolution image of Agenor shows its similarity to hi-res images of other convergence bands, including **Figs. 10.1 and 10.2d**.

The Evil Twin (**Fig. 10.5**) closely resembles Agenor (**Fig. 10.3**) but lies nearly diametrically opposite it on the globe (**Fig. 10.6**). Agenor runs westward from longitude 180° in the southern hemisphere, and its twin runs westward from longitude 0° in the northern hemisphere. Agenor, its Katreus component, and the Evil Twin are bright bands, with occasional narrow dark borders. The Evil Twin and Katreus each contain a large "island" of undisturbed older terrain. While quite similar to one another, Agenor (with Katreus) and its twin are quite different from other linear features on Europa.

Given this similarity alone, Agenor's twin became another candidate for classification as a convergence feature. Naturally, I could not resist referring to it as the Evil Twin of Agenor. Apparently, as soon as we had presented the significance of this feature at a conference, the nomenklatura of nomenclature in the International Astronomical Union rushed to give the thing a more respectable name. But they were too late. By the time I learned its official name *Corick Linea*, I had already published it as the *Evil Twin*.

What makes the Evil Twin special is that, unlike at the other convergence candidates, it has bordering plates with prominent markings, thus allowing for reconstruction. As mentioned earlier, Agenor and other convergence features have resisted reconstruction precisely because the terrain that

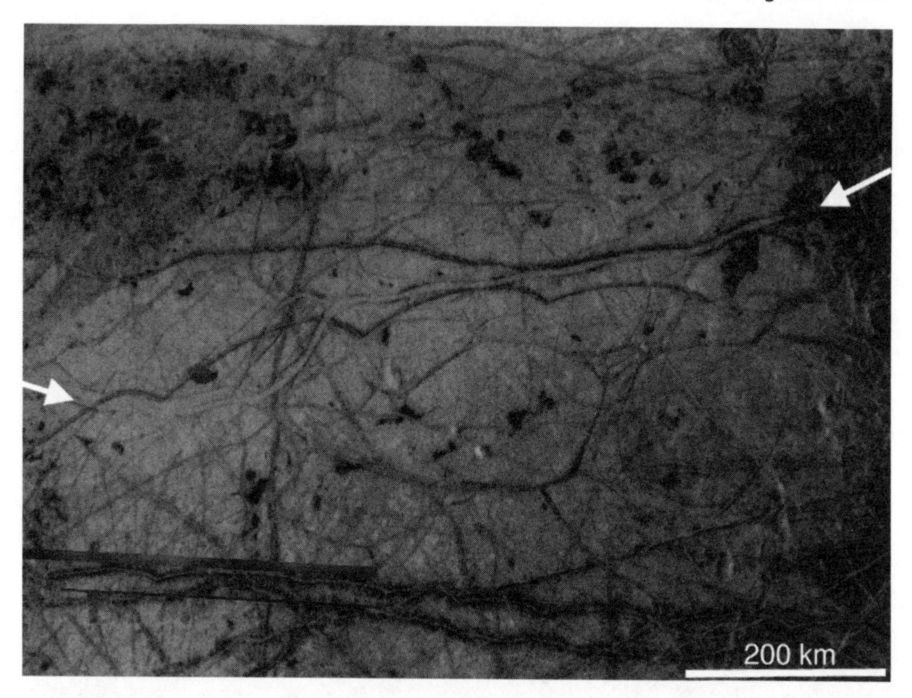

FIGURE 10.5: The Evil Twin of Agenor runs between the arrows, appearing as a light gray band similar to Agenor in Fig. 10.3. The twin includes an "island" (left of center) similar to that in Katreus (Fig. 10.3). The resolution here is about 1 km/ pixel. Note the circular and boxy crack patterns south of Agenor's twin, similar in shape and appearance to the crack patterns in the Wedges region, which are nearly diametrically opposite on the globe.

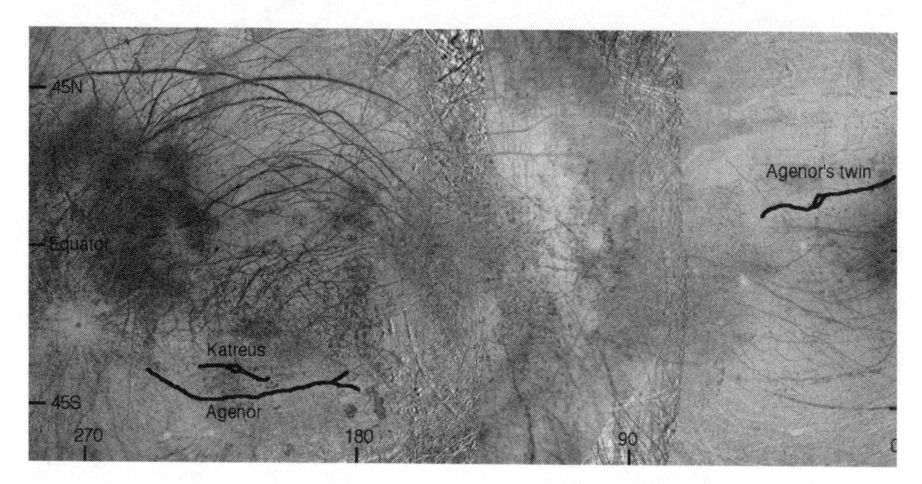

FIGURE 10.6: Locations of Agenor and its twin on a USGS global mosaic (Mercator projection). Agenor and its twin lie diametrically opposite one another on the globe.

previously existed between the current surface plates is now gone. The challenge is like the problem of finding the correct separation of two halves of a picture puzzle when you're missing the pieces that would fit between them. At the Evil Twin, reconstruction is largely guided by thick dark lines that obliquely cross the fault and also cross one another on the "island." They define a particular displacement of the neighboring terrain needed in order to bring each of these lineaments into alignment. To do the reconstruction yourself, purchase an extra copy of this book and cut up **Fig. 10.5.** (Only lines that predate the formation of the Evil Twin can be used for the reconstruction. More recent lines provide no clues about the pre-convergence past; They are easy to spot, because they cross right over the terrain scrunched together by the convergence process, within the band of the Evil Twin.) When you line up the dark crossing lines, you will get the arrangement that is shown in **Fig. 10.7.**

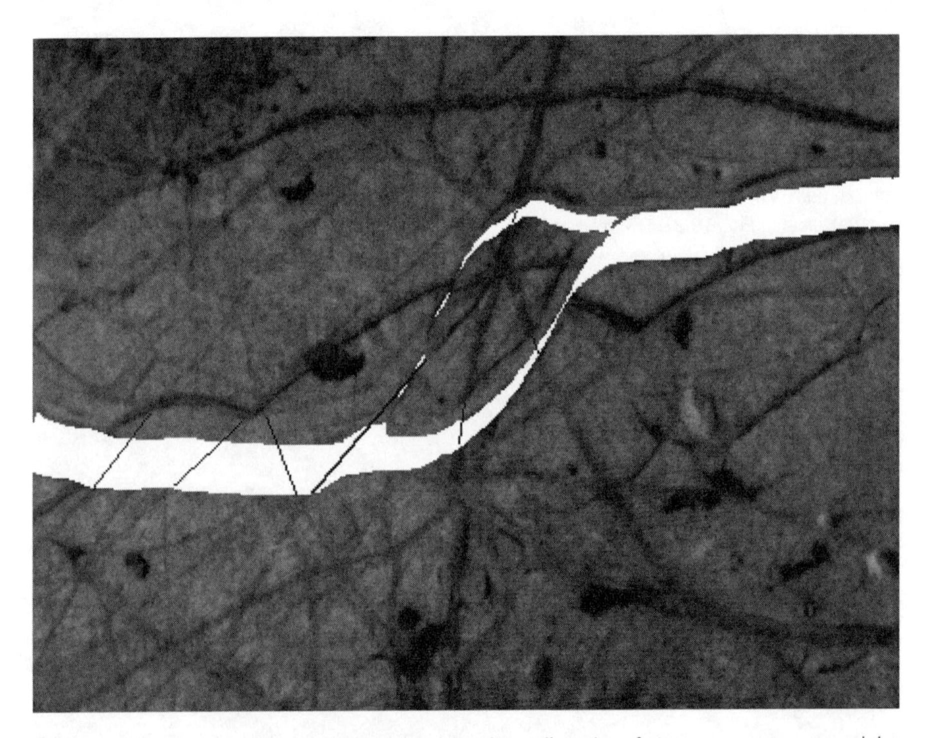

FIGURE 10.7: The area around the island in the Evil Twin of Agenor, reconstructed by realigning other lines that cross the area. Going back in time, a gap (shown in white) opens about 25 km wide, showing how much surface has been removed during convergence here. I kept the current bright band arbitrarily attached to the plate to its north. Minor linear features that are realigned corroborate the reconstruction, as marked. Dark lines that formed after the convergence (such as the prominent cycloid) are broken up by the reconstruction.

The reconstruction (**Fig. 10.7**), going back in time, reveals a gap 25 km wide. Over the 400-km length of this feature, a total of 10,000 km^2 of previous surface has been eliminated, a significant contribution to balancing the surface-area budget.

The Evil Twin of Agenor provides the first direct reconstruction of a convergence band on Europa, as well as strong supporting evidence that we have identified the general appearance of convergence features. We had already identified convergence based on neighboring strike-slip reconstructions, and at Agenor, where convergence had been inferred from its location relative to the dilational Wedges. The Evil Twin strengthens that hypothesis further, because it is so similar to Agenor and is demonstrably a convergence feature.

In terms of topography and structure, all of the convergence bands that we have found are subtle and subdued, just the opposite of what would be expected if the ice were very thick. As the Himalayas demonstrate on Earth, when thick plates collide, you expect the material to pile up as it is scrunched together in the collision. On Europa, such structures are not evident, even though substantial convergence has occurred. At Agenor's twin, for example, a zone of crust more than 25 km wide has been compressed into a feature only a few km wide, but the surface is fairly flat. If the ice layer were thick, such collisons would probably pile up mountains—not these nearly flat areas, which elsewhere I've dubbed the Himalayas of Europa. The lack of real mountains at convergence sites suggests that not much solid material was involved. That result is consistent with everything else we have been learning about Europa. The crust probably consists of only a thin layer of ice over liquid water.

The locations of Agenor and the Evil Twin nearly diametrically opposite one another on the globe also makes me wonder whether tidal processes have played a role in determining the character and locations of these convergence features in some way. After all, tidal stresses are symmetrical on opposite sides of the planet. Or, the locations may be simply coincidental.

There has been evidence for a quarter of a century, since Voyager, that new surface area has been created by dilation, and Galileo has shown that the new area has been continually created at widely distributed sites. Now, with identification of specific examples of convergence bands and of the characteristics that identify them, we are beginning to move toward a better understanding of the global surface-area budget.

Return to Astypalaea

Strike-slip faults had proven to be very important on Europa. They are common, and they provided a surprisingly large amount of information about the structure and processes of the little planet. After Randy Tufts discovered the archetypical example Astypalaea Linea ("The Fault") by going through old Voyager images, he wanted to go back to the real source and look more closely. Of course, that meant getting new images from Europe itself. And that in turn meant dealing with the imaging team.

Ordinarily I had little interest in the image-planning process—for two reasons. First, my feeling was it did not really matter where we looked, as long as it was on Europa. It would *all* be interesting, and sooner or later we were likely to get images with a fairly good sampling of the types of features and their broader context. As for my second reason not to participate: as discussed in earlier chapters, I was emphatically not welcome to work on image-planning. Who needed the grief? But then fate, in the form of Randy Tufts, intervened.

Randy loved "The Fault" on Europa nearly as much as "The Cave" in Arizona, so he convinced me to push for high-resolution images during a later orbit. Other team members (and their students and assistants) had their own agendas and targets in mind. For example, Ron Greeley, a volcanologist, was certain that some of the dark patches we had seen at low resolution must be volcanic lava flows. (He called them "cryo-volcanic," because the material involved would be cold water rather than molten rock.) Although similar dark patches had already proven to be chaotic terrain, Greeley was determined to continue his search by taking more high-resolution images of the things he thought were volcanic. In the end, high-resolution images of Astypalaea would prove to be some of the most revealing images of Europa, while the search for volcanic flows yielded only more typical chaotic terrain. But before any of this could happen, we needed to persuade the team to include Astypalaea in the high-resolution sequence.

As the head of our research group, I was already taking heat for stepping out of the imaging team's party line, so I had little political capital, and I was never very effective at these machinations in any case. Fortunately, Randy's long career in politics, social activism, and saving The Cave made him a very astute and effective operator. We also had in our group Paul Geissler, whose unassuming get-along demeanor was often our secret weapon in getting what we wanted.

The style of decision-making on NASA space missions is adversarial, even confrontational. The projects are divided into various scientific and engineering teams, whose jobs are to advocate for particular points of view. I suspect that this structure developed in response to engineering being a series of compromises between objectives and constraints, which involve understanding the best arguments, pro and con, for every issue. Whatever its origin, the structure and process yields decisions based on effective advocacy rather than on cooperation. The best space lawyer or politician gets his way.

I once saw things happen differently. In the early 1990s, I was at a meeting at JPL, and during a break I happened to notice a small subgroup in a conference room in the imaging-team area. There was a hot debate going on, so I decided to slip in and see what was up. At the time, Galileo, still on its way to Jupiter, was approaching Ida, one of two asteroids it encountered en route. There were about a dozen people in the room, and I realized that the group included, in addition to a few members of the imaging ("SSI") team, a few of our counterparts from the "NIMS" team, the scientists planning observations with the Near-Infrared Mapping Spectrometer instrument. NIMS was essentially a spectrograph, which would measure the composition of light in terms of its wavelengths—in other words, the infrared equivalent of color. Considering what a great spectrograph NIMS was, it was remarkable that it could actually take pictures, albeit a bit fuzzy, but with a precise wavelength spectrum for each of its pixels. That spectrum, it was hoped, would help determine what substances were mixed in with the surface ice.

The two teams facing off in the conference room had very different objectives. The imagers wanted to take great photos with their SSI camera, while the NIMSers wanted to optimize the details of the wavelength information. In practice, this meant that the NIMS people wanted the spacecraft to pass by Ida fairly close and with the sunlight shining straight down on the surface. The imagers wanted to stay far enough out to avoid smearing the images, and they wanted to take pictures illuminated obliquely, so that topography would be visible in high relief.

Both groups were in full-press mode, with enormous amounts of macho posturing and bravado. I was having fun watching these guys pretending to be a bunch of bull mooses, which was especially comical because what they looked like, and indeed what they were (except for me, of course), was a group of almost stereotypical science nerds. I wondered whether space missions tend to be run in such an adversarial fashion to give these guys a protected setting where they could act out their macho fantasies. Certainly, anyone from the real world would have found the scene hilarious. I kept quiet and watched as the performance went on and on.

Also in the room were a few JPL engineers who understood the practical issues and who would be responsible for carrying out whatever decision was reached. Most of them were women. While I was watching the rutting mooses pretending to butt heads, I was wondering what the women were thinking. Did they buy into the fantasy? Were they moose cows admiringly awaiting the outcome of the joust? Suddenly, one of them spoke up: "We are not going to listen to any more of this nonsense; you need to make up your minds and be done with this." The nerd studs froze in mid-rut. One of the more senior men stood up, drew a sketch on the marker board and said, "well I suppose we could fly-by on this side and pretty much satisfy most of the needs of each instrument." Everyone nodded in agreement and the meeting concluded within minutes.

The women had shown that the confrontational debate was a sham. To me, it confirmed the wastefulness of the adversarial format of mission development. If the same amount of energy that went into the debate had been put into cooperation, decision-making would have been far more effective.

But confrontation continued to be the default mode. When the possibility to revisit The Fault at high resolution became available, Randy persuaded me that we would need to fight to get it into the imaging sequence, and that it would be worth the trouble. By this time in the mission, these decisions were being made via telephone conferences. Early in the Galileo project, back in the 1970s and '80s, the imaging team had to meet every few months, usually at JPL in Pasadena. The plan had been that during the time that the spacecraft would be in orbit at Jupiter, we would spend most of our time in residence in Pasadena, so we would be able to work together continuously on updating the image-sequence plans and on image interpretation. By the time Galileo actually did get into orbit, however, the team members had gotten old, and no one wanted to spend months on end together in California. Fortunately, thanks to the delays in the mission, modern telecommunications and data distribution systems had finally been adopted that made it possible for people to butt heads at a distance, safely, from the comfort of their own offices.

To participate in the teleconference for planning the images to be taken during orbit E17, my research group and I gathered around a speakerphone to go to battle for high-resolution images of The Fault. As always, the picture budget was limited. Also, we were up against the hobby horses of some very powerful operators. Ron Greeley needed to take pictures of his non-existent lava flows. To make matters worse, there had been a computer glitch during orbit E16, which meant none of the planned images of Europa expected from that pass had been obtained, and naturally there were people who wanted to make up during E17 what they'd lost during E16. And as if these challenges were not already enough, Randy, with his political sixth sense, advised me that there was an undercurrent of jealousy and animosity from the team's powers directed toward me personally. So I kept a low profile—mostly. As head of our research group, early in the telecon I did make a few formal broad points about the significance of Astypalaea Linea, but then let Randy, with his deep understanding of The Fault and with his political sophistication carry the ball.

Ultimately, we did get a limited set of high-resolution frames at 42 m/pixel that spanned about a quarter of the length of The Fault, plus a few frames that covered the surrounding terrain at moderate resolution of 200 m/pixel for context. The region imaged in this sequence proved to be remarkable in that it shows most of the major types of tectonic features found on Europa, including strike-slip, dilation zones, double ridges and cycloids. It shows them at very high resolution, combined in interesting and unique ways.

Greg Hoppa recalls that one reason we were granted the high-resolution images of Astypalaea was that Pappalardo and Head wanted to see the cusps of the cycloidal ridges at high resolution. They believed that the morphology would prove their "linear diapirism" story of ridge-formation, which posits a process in which bodies of low-density ice (diapirs) rise to the surface of cracks, creating "lips" along their edges (as described in **Chapter 7**). To Head and Pappalardo, this theory, backed with supporting images, would disprove, or at the very least supplant, my group's tidal squeezing model. Once the images came down, however, we never heard any more from them on the subject. The appearance at the cusps is perfectly consistent with our ridge-building model, and very difficult to explain with linear diapirism.

***O *

Randy was right. The images of The Fault, when finally received and assembled, were stunning to look at, and incredibly informative.

Fig. 11.1a is a mosaic of the high-resolution images of Astypalaea Linea, which spectacularly superceded the old Voyager image of the region (**Fig. 9.1a**). (In **Fig. 9.1a** the fault runs from the upper right to the lower left; Here, it runs vertically from the top to the bottom, and only covers 1/4 of the length of the fault.) At the top of **Fig. 11.1a** lies the parallelogram seen near

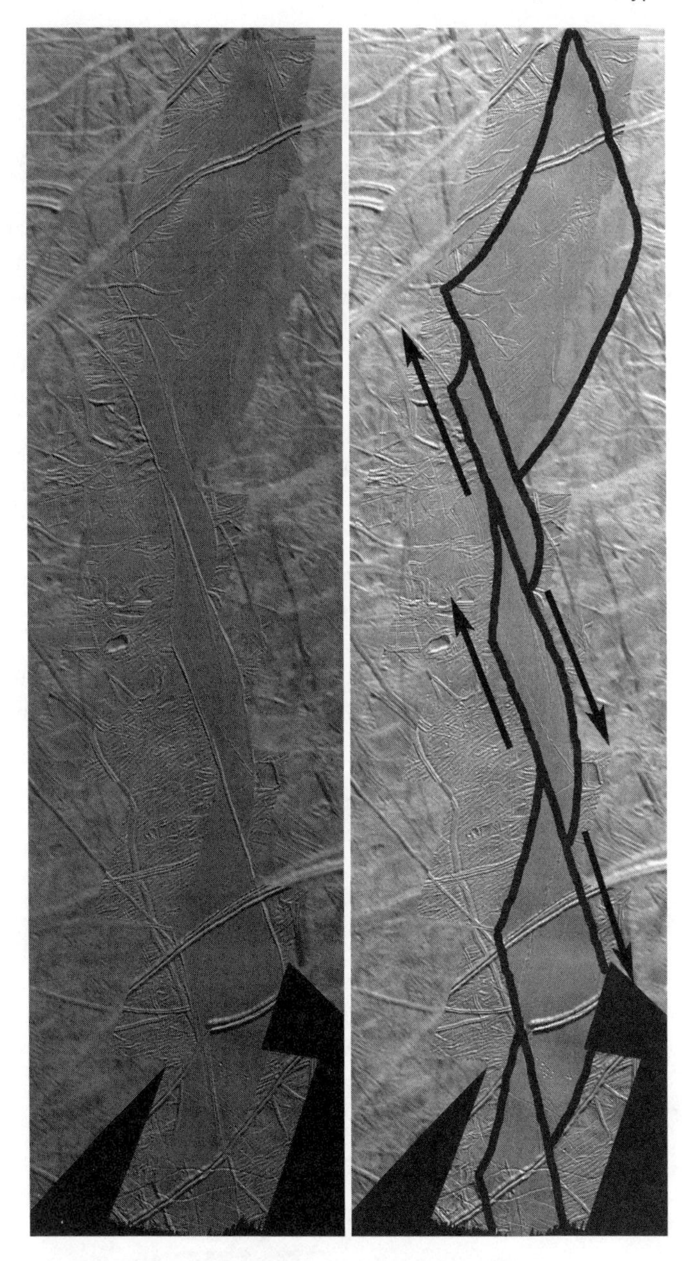

FIGURE 11.1: (a, left) A mosaic of the high-resolution images of Astypalaea covering about a quarter of its length (some 200 of 800 km), which was all we could get in the image-planning negotiations. The level of detail visible toward the middle (42 m/pixel) is not available toward the edges (where we only have 200 m/pixel data). The bottom of this mosaic, where we have no useful images, is blacked out. (b, right) The marked version shows the displacement of the neighboring terrain, with lines of shear and corresponding pull-apart areas that look just like other dilation bands.

the upper right end in the Voyager image. Near the bottom of **Fig. 11.1a,** Astypalaea is crossed by the cycloidal ridges that appear about a quarter of the way down the fault in the Voyager image.

An overlay diagramming the geometry of this quarter of Astypalaea is added to the mosaic in **Fig. 11.1b.** The overlay highlights a set of parallelogram-like pull-aparts. The high-resolution imaging revealed that there are several of these pull-aparts along the length of the fault, in addition to the one large parallelogram pull-apart that was identifiable in the Voyager image.

The high-resolution Galileo images in **Fig. 11.1** showed us that Astypalaea initiated as a somewhat cycloid-shaped fault. **Fig. 11.2** shows schematically how strike-slip along a cycloid can open a sequence of parallelogram-like openings. (They are not formally parallelograms, because one pair of sides is curved in each case.) On Europa, the slight diurnal working of the crack should cause parallel curved grooves within these pull-apart areas as sketched in **Fig. 11.2**, usually with a distinct center line where the newest material comes up. This schematic is similar to the actual geometry of Astypalaea seen in **Fig. 11.1.**

The mosaic of high-resolution images in **Fig. 11.1** runs about 8000 pixels from top to bottom. All the detail cannot appear on that single printed page, so to get a feeling for how rich it is, we need to look at separate blow-ups of different parts.

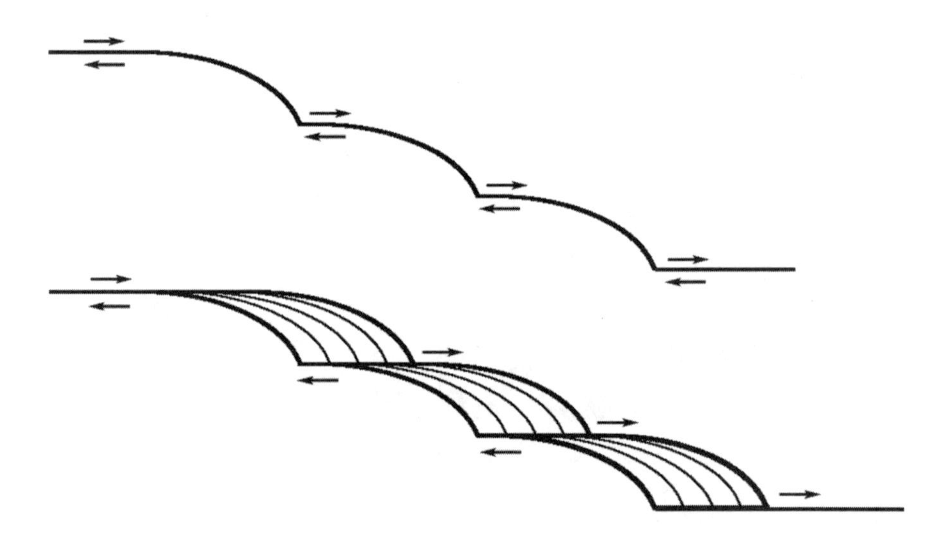

FIGURE 11.2: A schematic of the opening of pull-aparts when strike-slip (shear) displacement occurs along a cycloid-shaped crack. At the top is a cycloid-shaped crack; at the bottom is the geometry that can result from the displacement.

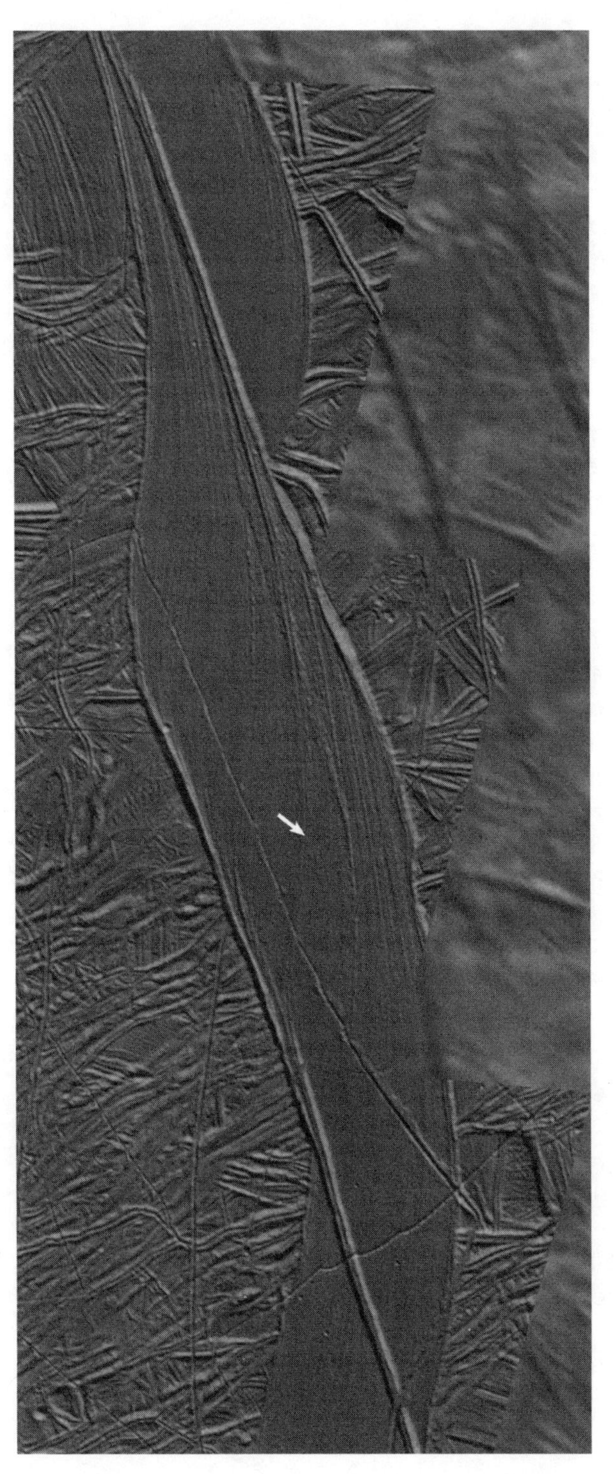

FIGURE 11.3: Enlargement of part of the high-resolution mosaic of Astypalaea in Fig. 11.1, showing detail of the pull-aparts and their similarity to other dilational bands, especially the fine furrows. The central groove is marked with a white arrow. Its path can be traced down the center of the pull-aparts and between the offset double ridges on the shear segments.

The pull-apart in **Fig. 11.3** has the typical look of a dilational band. The fine parallel furrows surround a central groove (marked by a white arrow). At the edges are low ridges that developed as the gap was just beginning to open. More pronounced ridges are also evident along the lines of shear that separate this parallelogram from the ones just above and below it. Because these boundaries were not being pulled directly open, diurnal tidal working was able to build ridges. (The crack to the left of the arrow formed after Astypalaea; it is interesting but not relevant here.) If we closed this parallelogram back up, the adjacent older terrain at the lower right would match up with the terrain at the upper left.

Like two trains passing in the night, the double ridges that developed along both sides of the shearing sections of the fault moved past one another. **Fig. 11.4** shows the shear zone between two parallelograms. (The top of **Fig. 11.4** overlaps **Fig. 11.3**.) The ridges sheared passed one another, so they are now double only in the middle of the shear zone. The central groove from **Fig. 11.3** cuts in between the two ridges, and then emerges from between them at the bottom to continue as the central groove in the next parallelogram.

I like the way all of these images also show tiny "secondary" craters, formed by bits of ice that were ejected when other, larger craters, perhaps Pwyll, were formed.

The next parallelogram (**Fig. 11.5**) shows other great details. Remember those long cycloids that had puzzled us since Voyager images showed them crossing over Astypalaea (**Figs. 6.3 or 9.1a**)? Here we see them close up, swooping across the pull-apart band. They turn out to be typical double ridges. By chance, we can even see the detailed appearance of a pointy cusp of one of the cycloids. Just to the left of the cusp, the double ridge seems to fade away; that is a lighting effect, because the illumination runs along the ridge there, minimizing shadows. Another cycloid near the top shows detail of a type of branching that seems to be fairly common. I will get back to cycloids in **Chapter 12**.

In **Fig. 11.5**, I increased the contrast somewhat to bring out the corrugations that I discussed in the **Chapter 10**, indicated by three or four dark zones that run perpendicular to the fine furrows. There are also sets of fine cracks crossing Astypalaea, which crudely correlate with the rises between the troughs. For example, one set of such cracks is seen just above the cusp of the cycloidal double ridge. This set is on one of the uplifted parts of the corrugation, but not at its high point. Moreover, the set of cracks at the bottom of **Fig. 11.5** seems to be at the lowest point of a trough. So this picture shows that the cracks do not fit well with the corrugations, one of several issues regarding these corrugations discussed in the previous chapter.

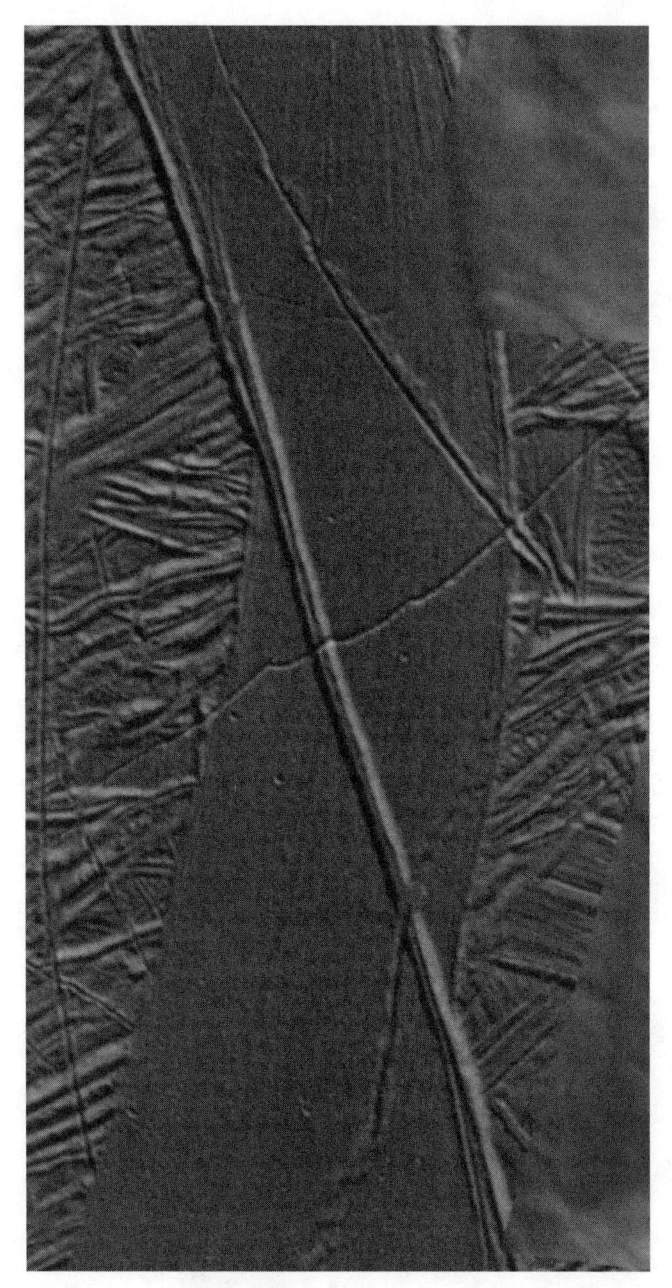

FIGURE 11.4: Enlargement of another part of the high-resolution mosaic of Astypalaea in **Fig. 11.1**, showing detail of one of the shear segments, with ridges passing like two trains in the night. The top part of this image overlaps **Fig. 11.3**.

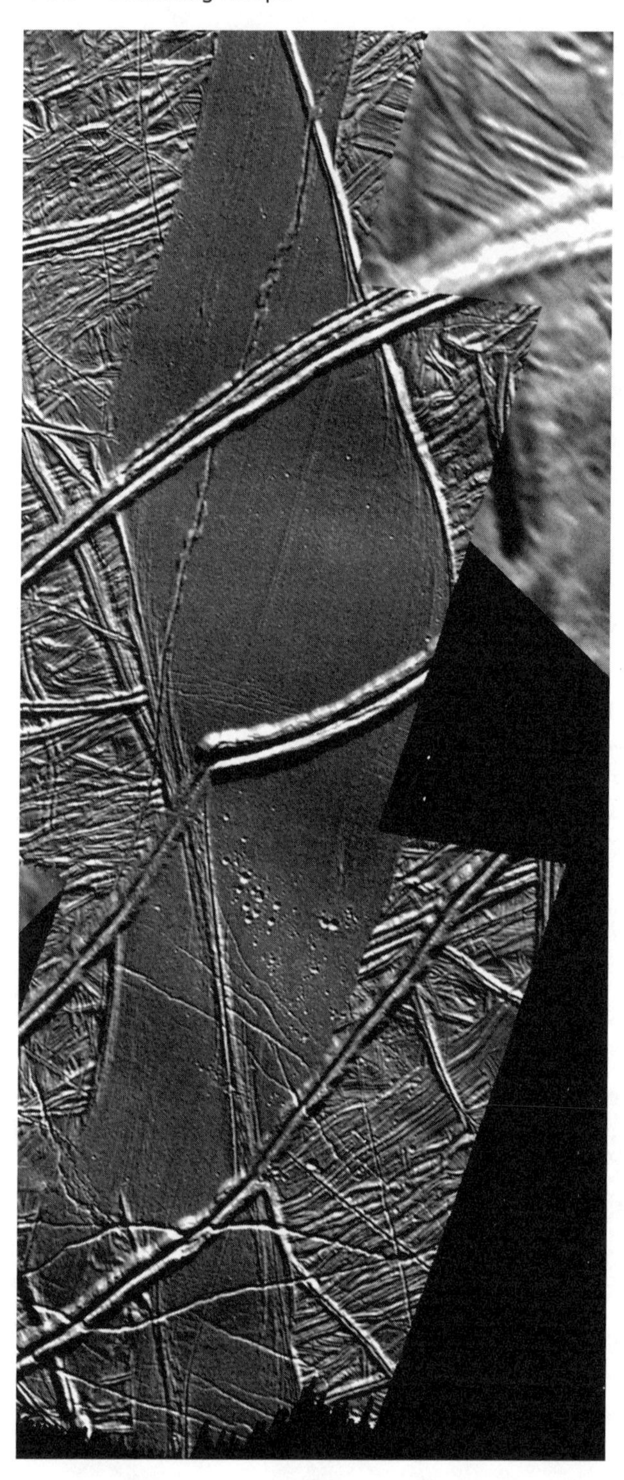

FIGURE 11.5: Enlargement of part of the high-resolution mosaic of Astypalaea in **Fig. 11.1**, showing detail toward the southern end of the portion imaged in this sequence. (For reference when looking at **Fig. 11.1**, look toward the bottom of the image, near the pointed and boxy black areas indicating insufficient data.) This area includes parts of the cycloid-shaped ridges that cross Astypalaea, as well as the corrugations (25-km wavelength) in the pull-apart zone. The corrugations run roughly perpendicular to the fine furrows.

Astypalaea Linea consolidated everything that we had seen on Europa. The 40-km offset on an 800-km-long fault confirms the mobility of the crust over a low-viscosity decoupling layer. The right-lateral displacement fits the tidal walking theory. The pull-apart segments have exactly the same morphology as typical dilation bands, only here we do understand the driving mechanism. The ridges along the shear zones, passing like trains on opposite tracks, confirm the penetration to liquid that both produced the ridges and allowed the strike-slip displacement. Everything we see at Astypalaea is consistent with thin mobile ice over liquid water.

Cycloids 12

AMONG the weirdest things revealed about Europa by Voyager in 1979 were the cycloid-shaped ridges, especially those crossing over Astypalaea and points east, where fortuitous lighting showed them well. We already saw these Voyager images in **Figs. 6.3 and 9.1a**, as well as a close-up of the same cycloids in **Fig. 11.5**. The International Astronomical Union, the same organization that recently decreed that Pluto cannot be called a planet, in its collective wisdom decided to call these features "flexi," and gave official names to some of them. But everyone calls them cycloids, in reference to their geometric shape—chains of arcs connected at pointy cusps. IAU nomenclature is random and inexplicable, the inevitable result of a huge amount of discussion over long dinners, but in this case it was harmless. (In a later chapter I'll discuss how some IAU-sanctioned misnomers actually promoted misunderstanding about Europa.)

Galileo images showed cycloidal cracks everywhere on Europa. Like most cracks, they usually show up as double ridges, but can also develop into strike-slip faults (like Astypalaea Linea) or dilation bands (like **Fig. 8.3** or most of the bands in the Wedges region). They may also remain as hard-to-spot simple cracks. The beautiful cycloids that formed across Astypalaea (**Fig. 6.3**) are typical in form and size: Each is ~100 km long, and chains often comprise a dozen or more connected arcs running across a thousand kilometers or more.

Cycloids are ubiquitous on Europa. You could scan almost any of the Galileo images and find beautiful examples. Some of my favorites run roughly north-south in the northern leading hemisphere (**Fig. 12.1**). Along much of the length of these cracks, double ridges have formed. Elsewhere the cracks have opened, forming dilational bands.

During the long interval between Voyager and Galileo, my student Mike Nolan and I had considered the possibility that the cycloids formed in a way similar to the island arcs on Earth. (A chain of arcs surrounds the Pacific Ocean, for example.) On Earth, those features are sites of convergence and subduction, where one plate dipped down under another. In the sea floor

FIGURE 12.1: These cycloidal ridges (including some showing dilation) are in the northern hemisphere. This Regional Mapping image was taken during orbit E15 at 230 m/pixel, but it is foreshortened by the viewing angle toward the north pole.

there is usually a deep trench on one side of the island arc. If the same thing had happened on Europa, we would have expected to see something similar as one plate is pushed down under the other. But our idea didn't pan out. Nothing like that is found along the Europan cycloids. The ridges are as symmetrical as ridges along any other cracks. So the cycloids probably formed as simple tension cracks, like most of the other lineaments on Europa.

To me the cycloids, unique to Europa, were so striking and so common there that they seemed to be a key to the very character of the place. I tried to keep them as a high-priority issue for my research students. Fortunately, my students Randy Tufts and Greg Hoppa agreed that the cycloids were worth their attention.

***○ *

Remember, in broad patterns the lineaments of Europa correlated with tidal tension. And Randy, as the structural geologist in my research group, kept emphasizing that ice tends to fail in tension, and is less likely to crack under the stress of shear or compression. Randy developed his ideas about tensile cracking with an intuitive and graphic approach. He drew lots of little pictures. And he was giving a lot of thought to the cycloids.

He shared a temporary cubicle in my lab with Greg Hoppa, and as Galileo images became available, he printed them on sheets of standard-size paper, scotch-taping them into huge mosaics covering every wall and partition in sight. Greg and Randy discussed these images constantly. At the same time, Greg was producing the global-stress diagrams (like **Fig. 5.1**), which, in the format that I had designed, gave an intuitive, graphic picture of how tidal stress changed over time.

Randy often came to our weekly research group meetings with notebook sketches of how the tectonic structures of Europa might have evolved. We had seen sketches used by other geologists associated with the Galileo imaging team, and often those drawings were the sole basis for published interpretations of Europa's geology. The philosophy of those authors seemed to be that, if you could draw it, it must be true. The physicists in my own group called that approach cartoon science. It reminded me of scenes from Road Runner cartoons in which Wile E. Coyote runs off a cliff but stays aloft until he realizes there is nothing holding him up. Randy assured me that the problem was not that geology as a discipline was inadequate, just that we were seeing a lot of bad geology. In any case, with so many physics-types in my group, Randy was not likely to sell any cartoon science unless it could be supported by quantitative analysis.

One week Randy showed us a series of sketches of a crack propagating across the surface of Europa, with the direction and magnitude of the stress changing according to Greg's calculations. The result was a smoothly curving trajectory of cracks. Naturally, I was skeptical, but Greg had been in on the development, and the stresses were based on quantitative theory. Greg and Randy worked up the model in detail. To the delight and excitement of everyone in our group, the patterns that follow from propagation of tensile cracks over the course of several Europan days follow exactly the patterns that were observed on Europa.

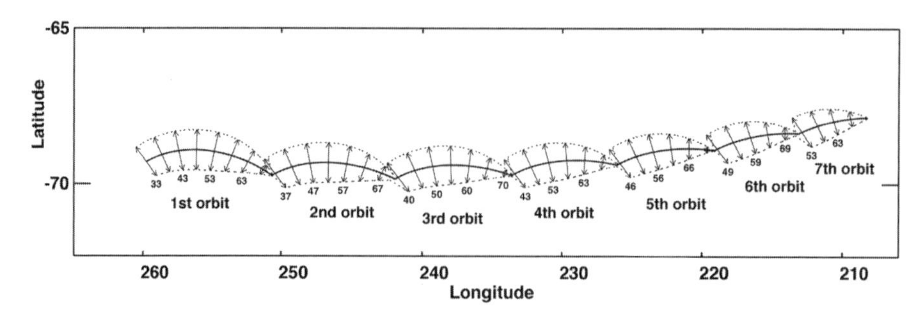

FIGURE 12.2: Propagation of a cycloid from an initial cracking site (at left) in the far south of Europa. The double-headed arrows show the tidal tension at the tip of the propagating crack, with a number showing the time-of-day in hours (out of the 85-hour Europa day). At time 33 on the first day, the tidal stress has become great enough to initiate tensile cracking. As time goes on, and the crack propagates to the east, the tidal stress (arrows) changes direction, resulting in curved cracking. In intervals when the stress is low (seen here initially when the crack reaches the end of the first arc), crack propagation comes to a halt for several hours. Then, after the tensile stress becomes stronger (and changes direction again), the process starts all over again, resulting in the formation of a cusp and then the next arc.

Consider the example in **Fig. 12.2.** Suppose cracking starts at latitude 69°S and longitude 260°W at a time when the diurnal tidal stress is in tension and exceeds the strength of the ice. The crack then propagates across the surface (toward the right in the case shown in **Fig. 12.2**), while at the same time Europa moves in its orbit so that the tides change. So at the same time as the crack propagates eastward, the diurnal tidal stress changes direction and magnitude, as shown in the diagram by the double-headed arrows. The strength and direction of the tension are represented by the length and direction of the double arrows, and the number shows the time (in hours) in Europa's 85-hour day. As the crack propagates, its new extensions always must form perpendicular to the tension, so the changing stress direction causes the crack to curve. As time goes on, the tension decreases, as well as changing direction. So eventually, in this case about 35 hours after the crack initiates, the tension decreases enough that the crack stops propagating for a while.

Then, roughly one Europan day after the crack first began to propagate, the tidal tension grows enough for the crack to start up again, and propagate further. (This happens *roughly* one day later because a single complete diurnal rotation of Europa is nearly synchronous with a single orbital trip around Jupiter, but the propagation point is now about 100 km east of the starting point.) At this time, the direction of the tension is similar to what it had been a day earlier. But it is quite different in its direction from what it had been just before the crack went to sleep hours earlier. So the new direction results in a cusp—a sharp turn that is a key part of the distinctive cycloid pattern.

This sequence repeats itself, each day yielding a new arc and cusp. A typical cycloid of ten or twelve arcs takes that many Europan days to form, or about one Earth month. The details of the diurnal change in tensile stress vary with each day, as the crack propagates to locations hundreds of kilometers away from the starting point, so the chain may gradually change direction. The size and shapes of arcs may change along the chain as well.

This beautiful dance between the orbitally generated tides and the response of the ice creates the almost unnatural cycloid shapes that we see all over Europa. In fact, this theoretical model has proven successful at explaining the shapes and orientations of cycloids all over Europa. It even explains the curvy shapes of the cracks in the Wedges region. To be sure, to make things fit in detail we have to make plenty of assumptions about things like the strength of the ice and the speed of crack propagation in it. But none of the necessary assumptions are unreasonable. The tidal model has been widely accepted as the basic explanation for the cycloidal crack patterns.

The theory of cycloidal crack patterns would have been even more satisfying if we knew the properties of the ice, and could show that they led to exactly what we see on Europa. The best we can do is make reasonable estimates and see whether they fit the observations. The theory would have been unconvincing if a different set of parameter values was needed to fit each cycloid. What saves it is that most of the cyclical features on Europa can be explained with very similar values for the parameters.

The required properties of the ice seem reasonable. The theory works well using strength and elasticity values for the ice that are similar to what we assumed for our other calculations of tidal stress effects. The speed of propagation inferred for each cycloidal crack is comparable to a typical human walking speed. The Discovery Channel on television showed footage of me walking across the Arctic ice sheet next to an animated propagating crack, so evidently this speed seems reasonable. The consistency of the inferred parameter values for most cycloidal features offers some assurance that the model is a good representation of the process that controls many of the important crack patterns on Europa.

Not all cracks follow cycloidal patterns, however. One reason may be that the surface has been so sliced and diced and rearranged that older crack patterns are hard to recognize. Indeed, many tectonic lineaments are often slightly curved even if they are too short to show the full cycloidal shape. Most recognizable cycloids are fairly recent; the ones that cross over Astypalaea are good examples. Furthermore, we know from the existence of chaotic terrain that from time to time and place to place, the properties of the ice must have been different from average. Cracks crossing such places would be irregular. Full-fledged cycloids required the ice to have been fairly uniform over a huge area for at least a month. Most significantly, the fact that all cycloids seem to have formed in similar ice conditions means that, whenever the surface has been uniform over a large region, it has been in a condition similar to what existed at all other times and in all other places where cycloids have formed.

***○ *

As we thought about cycloid formation, we realized that it could help us learn more about Europa's rotation. Remember, tidal theory had made nonsynchronous rotation seem plausible. We also knew from Greg Hoppa's comparison of Voyager and Galileo images that the rotation (relative to the direction of Jupiter) had to be slow, with a period greater than 12,000 years. And both the amounts of strike-slip displacement and the sizes of the largest ridges suggested that cracks open to the ocean froze shut after at least a few tens of thousands of years, consistent with nonsynchronous rotation with a period of ~100,000 years.

Cycloids gave us another way to study Europa's past rotation. For any observed cycloid, its specific shape and the tidal theory told us where it could have formed, relative to the direction of Jupiter. And by studying the order in which these features formed, based on how younger ones cut across older ones, we can begin to sort out the rotational history. As discussed in **Chapter 6**, Greg's pilot demonstration of this technique yielded a rotation rate of ~50,000 years, consistent with other estimates. There are plenty of other cycloids so, as we continue to characterize their appearance and refine the tidal theory, there is the potential for learning a great deal more about the history of Europa.

In any case, most of the features with recognizable cycloidal patterns are very recent in the recorded geological history of Europa. Older faults have been sliced and diced by subsequent tectonics to the point where they are no longer recognizable. Though the recorded geological history of Europa goes back a few tens of millions of years, because of the continual resurfacing the useful record of cycloids goes back only a few million years at most. Yet even in that short time significant and complex reorientation of the ice shell has occurred, according to the evidence for polar wander and nonsynchronous rotation.

Our understanding of the origins of the cycloidal patterns provided the first definitive evidence that Europa did indeed have a global ocean. (At least, the *New York Times* and the BBC called it the "strongest evidence yet" and the "most convincing evidence yet," respectively.) The strong correlation between theory and observation showed that tides were driving the cracks, and only with a global ocean could there be adequate stress. Randy and Greg's discovery was front-page news across the US. My own feeling was that the general characteristics of the tectonic and chaotic terrains were already convincing enough evidence for the ocean. But after we explained the cycloids, even the most conservative scientists acknowledged that there had to have been an ocean.

But conservative hold-outs would not agree that there necessarily is an ocean *now*. The way I looked at it, there had still been an ocean during at least the past 0.1 percent of the age of the solar system while the cycloids were cracking. And the geological activity on Europa had been rapid but fairly constant over at least the past 1 percent of the age of the solar system. So it seemed strange to think that the ocean would have suddenly (on the cosmic scale) frozen just in time for Galileo. It seemed obvious to me from the geological record that the ocean must still be there.

Over the next couple of years, after we had developed the cycloid model, additional evidence for a currently existing ocean accumulated from the magnetometer on Galileo. This instrument had measured changes in the magnetic field around Jupiter as the spacecraft swooped past Europa. These changes were consistent with what would be expected if Europa had a global shell of electrically conducting material, such as salt water, just below its surface. According to the theory, as Europa orbited through Jupiter's magnetic field, electrical currents would be induced in the conducting shell, in turn generating a magnetic field that would add to the Jovian field. The changes detected by Galileo's magnetometer fit that model. This result complements the strong evidence from the cycloids that an ocean was present while the current surface was being created, indicating that the ocean still existed as recently as the late twentieth century.

The cycloids remain a critical part of the evidence, however, for a couple of reasons. First, the magnetometer results depend on acceptance of the conducting ocean as the unique way of explaining the variations in the detected field. That model is simple and elegant, so Ockham's Razor gives it great credence. Whether any other adequate models might work is an open question. The existence of the cycloids, therefore, provides essential supplementary evidence. Second, the fact that cycloids formed so recently compared with the age of Europa would make it very implausible that the ocean would not still exist today. So cycloids, as well as the magnetometer data, do relate to the current state of Europa. Both the cycloids and the magnetic field model offer good evidence for the ocean. With both considered together, the case is very strong.

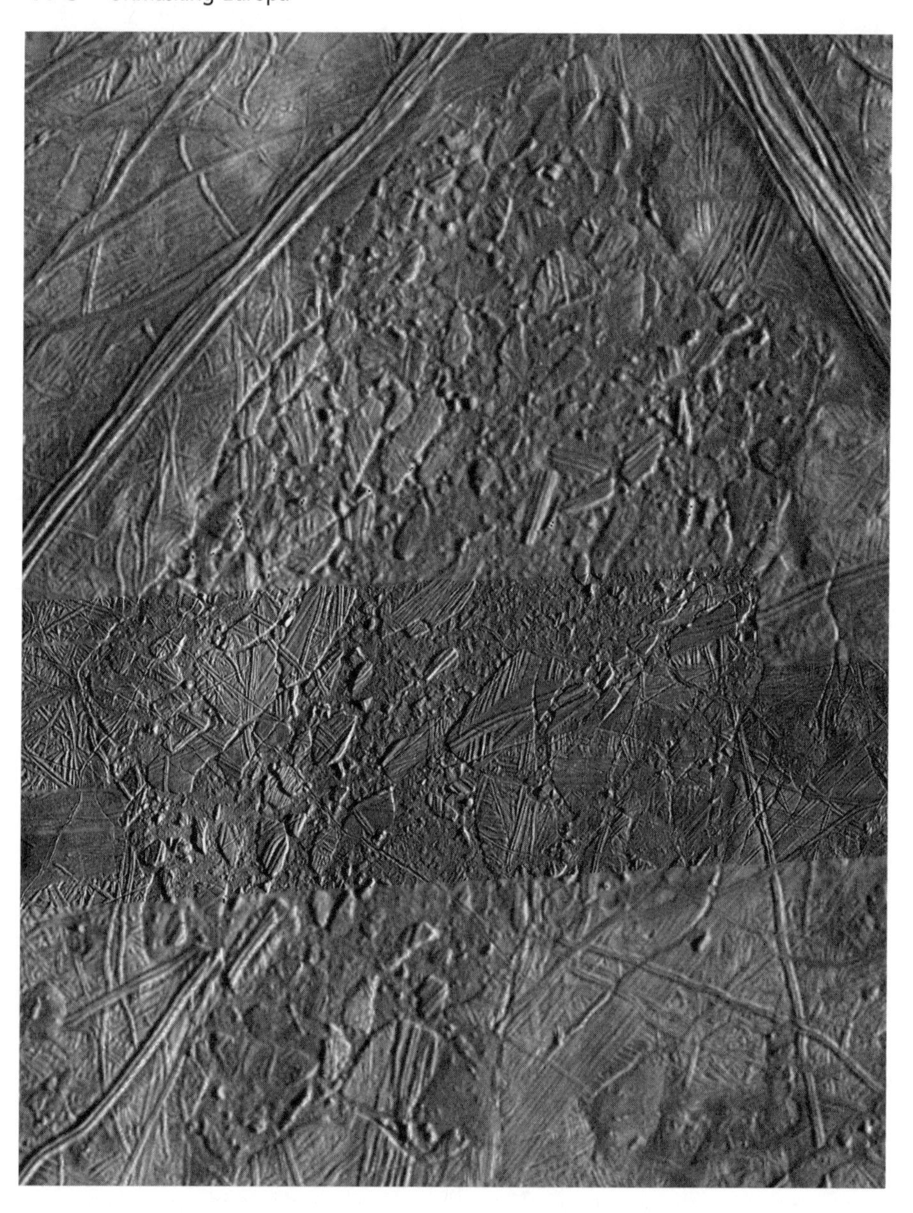

FIGURE 13.1a: Conamara Chaos imaged during Galileo's E6 orbit. The chaos area is about 80 km wide. This mosaic of images (mostly ~200 m/pixel) includes a set of higher-resolution (54 m/pixel) images (also from orbit E6) pasted across the middle. The location of this figure within the larger area shown in **Fig. 2.7** is shown in part (b).

Chaos

On Earth, when the icy surface of a northern lake or the Arctic ice cap breaks up, rafts of ice float around on patches of open water. These free-floating rafts may be portions of the ice crust that were thick enough to resist melting, or they may be large pieces of ice that broke off from the intact crust surrounding the melted zone, then drifted free on the water. In either case, if the weather turns cold again, the rafts freeze in place, with a matrix of refrozen water between them.

When we obtained our first decent images of Conamara chaos in February 1997, we saw a surface that is indistinguishable from such terrestrial melt-through sites (**Fig. 13.1**). Large rafts of ice display portions of the previous surface, densely ridged tectonic terrain like what one sees on Conamara's surroundings.

The rafts in these images can more or less be put back together like pieces of a jigsaw puzzle, allowing re-assembly of their original positions relative to the surrounding terrain and to one another (**Fig. 13.2**). To do this, you first erase the recently refrozen matrix between the rafts (making it white), then cut out the rafts, and finally arrange them in positions that match up and align the features on the various rafts and on the surrounding terrain, as

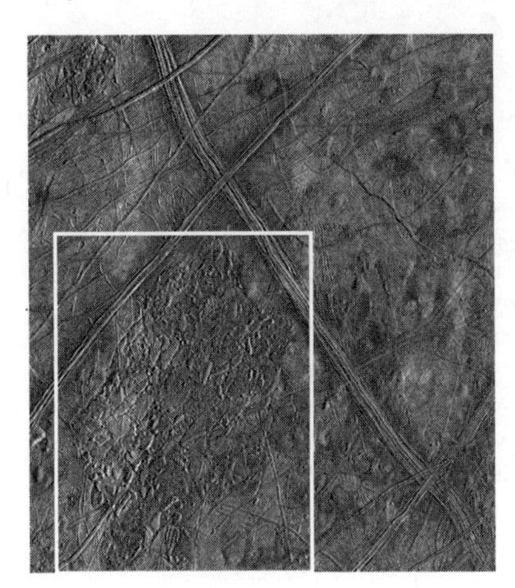

FIGURE 13.1b: The location of Fig. 13.1a is shown by the white rectangle within this thumbnail version of Fig. 2.7.

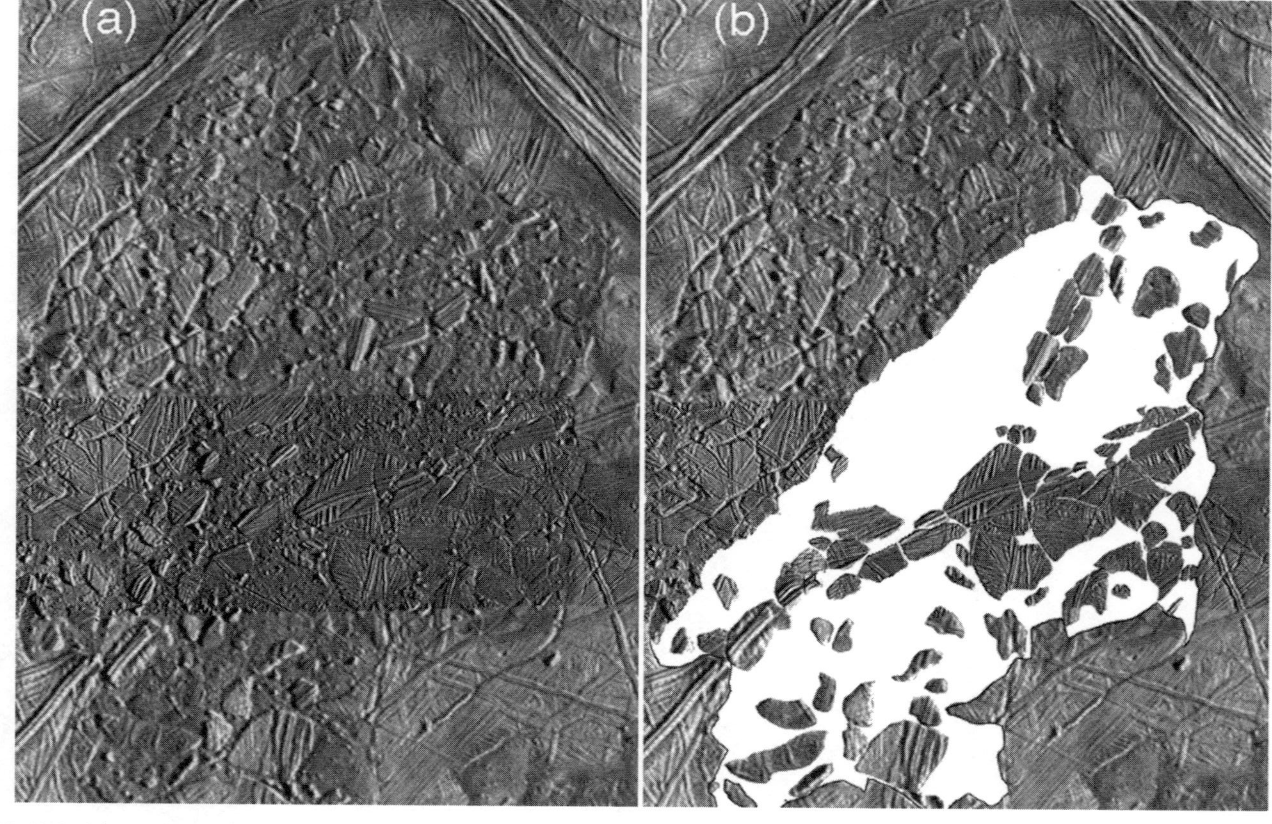

FIGURE 13.2: (a) A mosaic of Conamara Chaos, similar to **Fig. 13.1**, but geometrically corrected to give a view as if looking straight down on the region. (b) Repositioning of many of Conamara's rafts (after erasing to white the matrix between the rafts) shows them to be the remains of portions of the previous tectonic terrain where major ridge systems crossed. Chains of rafts contain nearly all of the double ridge that runs from the lower left across to the middle right. The rafts to the northwest have not been reconstructed here.

shown in **Fig. 13.2b**. Comparing these reconstructed original positions (**Fig. 13.2b**) with the current locations of the rafts (**Fig. 13.2a**), we can see how much they have moved. (In **Fig, 13.2** we have only reconstructed the southeastern portion of Conamara. The rafts in the northwestern half, and the matrix between them, have been left in place because they evidently have not moved very far.)

Often in reconstructing a chaos area, the placement of the rafts back into their original locations results in chains that reconnect the ridges that formerly crossed the area. For example, the prominent double ridge that approaches Conamara from the southwest (from the lower left corner in **Fig. 13.2b**) can be traced across the realigned rafts, curving slightly to where it is reconnected to its continuation east of Conamara (at the center right side of **Fig. 13.2b**.) Without the reconstruction, the current appearance (**Fig. 13.2a**) would make it impossible to recognize that the segments of double ridge at the lower left and at the middle right were originally formed together as parts of the same double ridge.

Toward the westward side of Conamara Chaos, the rafts have moved very little away from the banks of the opening, as evidenced by the fact that they still fit together quite well, with little new matrix between them. Along the southeastern edge of the opening, cracks in the adjacent terrain suggest that incipient rafts have also stayed in place. Aside from the fact that they have not moved, there may be no other difference between them and rafts that have drifted into the opening.

The relationships among the rafts and the matrix are clarified further in the enlargement in **Fig. 13.3**. Here we can see in more detail most of the rafts involved in the reconstruction (**Fig. 13.2b**) of the ridge that crossed this area. We can see that the lumpy, bumpy matrix between the rafts appears to consist of small bits of previous, unmelted crust too small to show intact features of the pre-chaotic terrain. As noted in **Chapter 2**, new cracks and ridges have already begun to form across this refrozen terrain, beginning the transformation of the chaotic region back to tectonic.

East of Conamara Chaos, on the right side of **Fig. 13.3**, we see small, distinct, irregularly shaped patches of chaotic terrain. Each has the usual characteristics of chaos, including small rafts and the lumpy bumpy matrix. On the other hand, the terrain between these small patches of chaos and the main part of Conamara has been broken somewhat by cracks. Aside from having not been displaced, this terrain could be considered to be a set of large rafts lying between the exposed matrix in Conamara and in the small patches to the east. Viewed that way, the small patches are actually part of the Conamara structure, and they likely formed as part of the same disruption event.

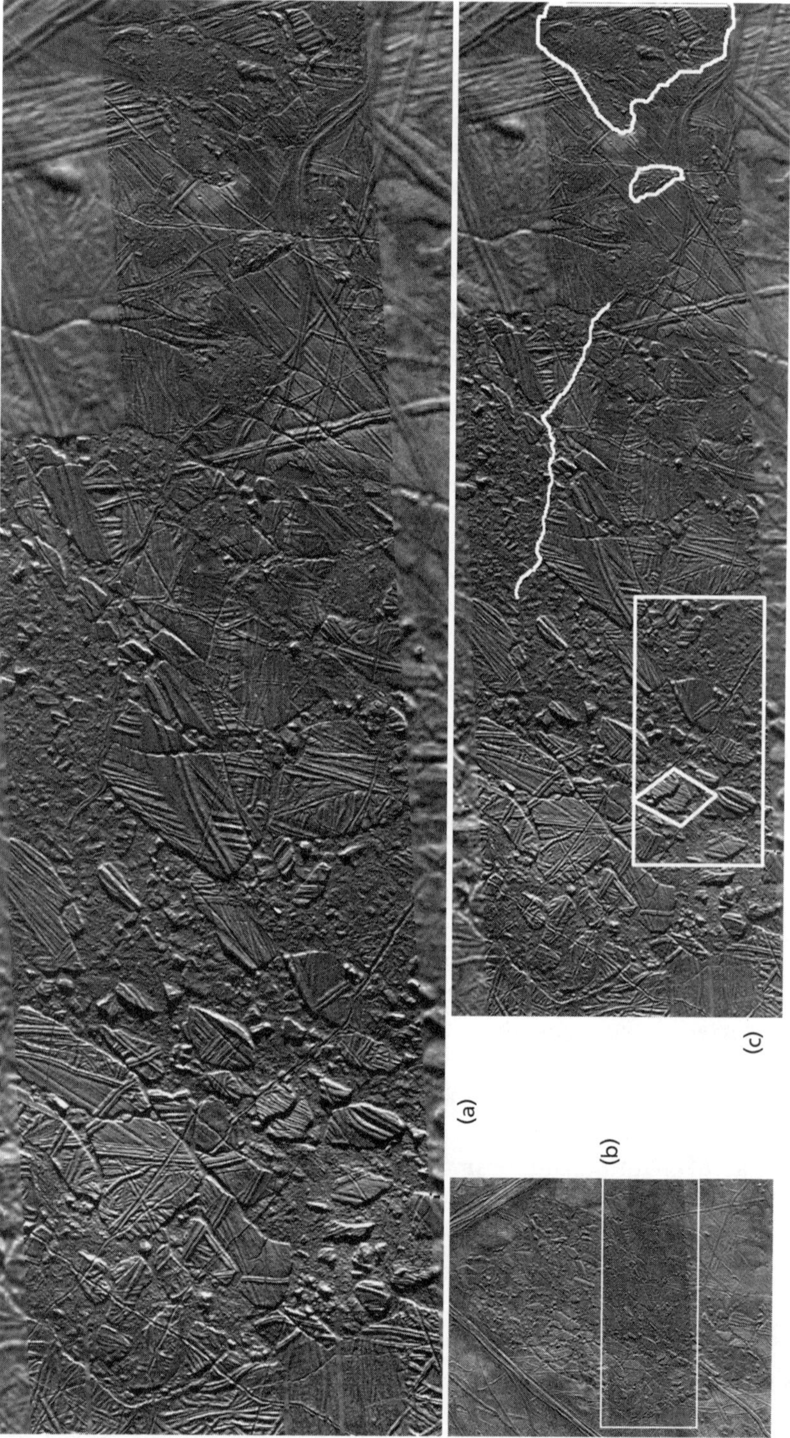

FIGURE 13.3: (a) An enlargement of the higher-resolution part of the Conamara Chaos mosaic. Some sections with lower resolution, appearing relatively fuzzy, are seen mostly at the top and bottom. (b) A thumbnail version of **Fig. 13.1a** shows the part of that figure included in (a) here. The higher resolution in (a) shows details of the rafts and the lumpy, bumpy matrix between them. Part (c) is a version of (a) with various features outlined: Separate patches of chaos lie east of the main part of Conamara near the right side of (c). A relatively recent ridge that crosses the chaos is marked by a squiggly line in (c). The rectangle and diamond within (c) show locations of even-higher-resolution images shown in **Figs. 13.4** and **13.5**.

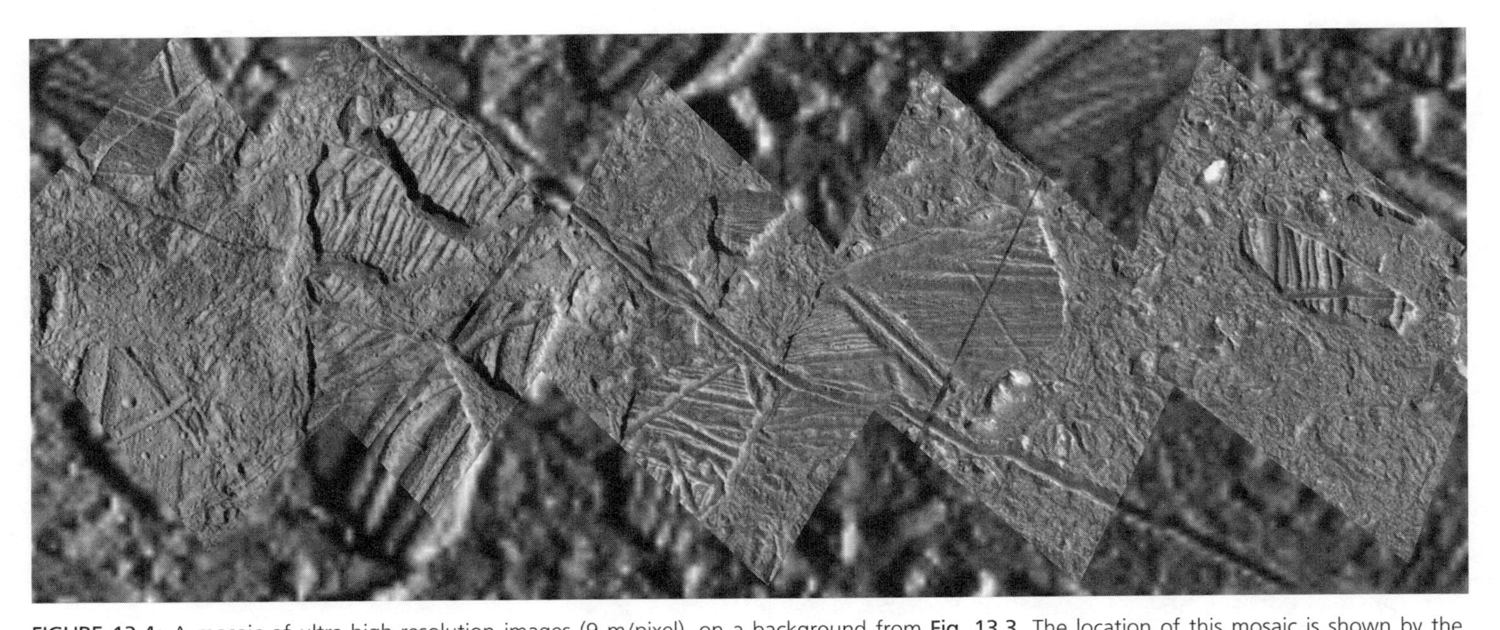

FIGURE 13.4: A mosaic of ultra-high-resolution images (9 m/pixel), on a background from **Fig. 13.3**. The location of this mosaic is shown by the rectangle in **Fig. 13.3c**. The background here, which seemed so sharp and detailed in **Fig. 13.3a** (compared with **Fig. 13.1**), appears fuzzy relative to the super-high-resolution portions. Like most of the mosaics used in this book, this one was assembled by student assistants at the University of Arizona's Planetary Imaging Research Lab. This mosaic spans about 30 km from east to west.

FIGURE 13.5: Part of the ultra-high-resolution image set shown in **Fig. 13.4**, but here in its original viewing geometry: an oblique view, as if from the window of an airliner. This amazing image (only about 5 km across) shows the depth of the drop-off from the tops of the rafts to the refrozen matrix below. Cliffs at the edges of the rafts cast shadows toward the left onto the matrix below. At the bases of the cliffs facing us are fans of spread-out debris. In the foreground, a fine crack has formed across the matrix. A few tiny craters are visible on the rafts.

In December 1997, during orbit E12, a set of several ultra-high-resolution images (9 m/pixel) were taken of an area within **Fig. 13.3.** In **Fig. 13.4,** these high-res images are set in place in the background of the earlier images. In fantastic detail they show the rafts and matrix between them. Toward the west end of this mosaic is a raft covered by parallel ridges which has cracked in two. An oblique view, looking into the crack as if from an airliner's window is shown in **Fig. 13.5.** It shows the detail of the cliffs dropping from the sides of the rafts down to the refrozen matrix below.

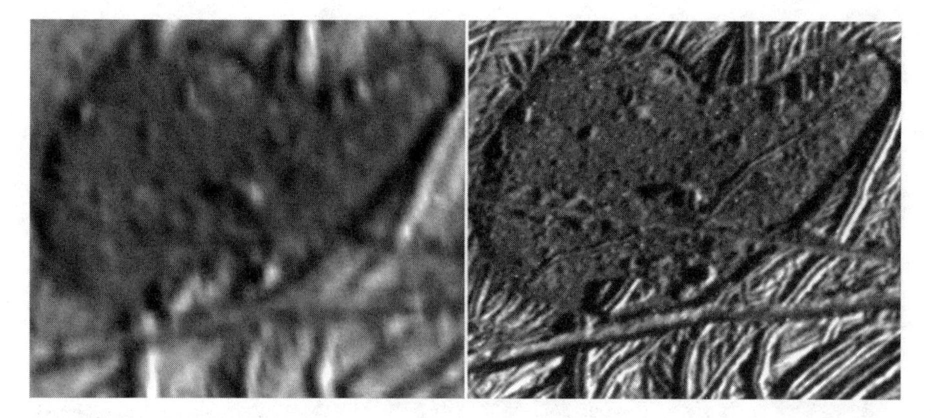

FIGURE 13.6: The "mini-mitten," a small patch of chaotic terrain. This image is about 18 km across. At left is a view at 230 m/pixel from a Regional-Mapping image. At the right is the same feature at 64 m/pixel. At even lower resolution (> 1 km/pixel) this feature would appear as one of the dark spots for which the name "lenticulae" was invented (e.g. **Fig. 2.6**).

The general character of Conamara is typical of fairly fresh chaotic terrain—with rafts detached from, but still recognizably corresponding to, surrounding terrain, and refrozen matrix just beginning to show its own tectonic features. There is, however, no typical size for a patch of chaos. The example shown in **Fig. 13.6** is much smaller—about 1/6 as wide as Conamara. The identifying details of chaotic terrain can be difficult to recognize if the patch of chaos is small and the resolution is marginal (**Fig. 13.6, left**). However, with adequate resolution (**Fig. 13.6, right**), even such a small patch can show all of the identifying characteristics of chaotic terrain: rafts displaying parts of the previous tectonic surface, especially where the crust was thicker with ridges; the lumpy bumpy matrix; and cracks and ridges formed across the chaos since the matrix refroze.

In general, these identifying characteristics of chaos scale down with the size of the patch of chaotic terrain. For example, the small chaos in **Fig. 13.6** contains the same features as Conamara, but the rafts are smaller and the matrix texture is finer, more or less in proportion to the size of the chaos.

As a result, patches of chaos reach a limit of "recognizability" if they get too small relative to the pixel size in a given image. For another example, consider the small patches of chaotic terrain in **Fig. 13.7**, each a few kilometers wide and surrounded by fairly intact tectonic terrain. In the section of a Regional Mapping image at 230 m/pixel (on the left), the features are nearly unrecognizable. But at 34 m/pixel (right), the characteristics of chaotic terrain are clearly visible. So these small chaos patches are hard to recognize unless you have high enough resolution. (This observational bias

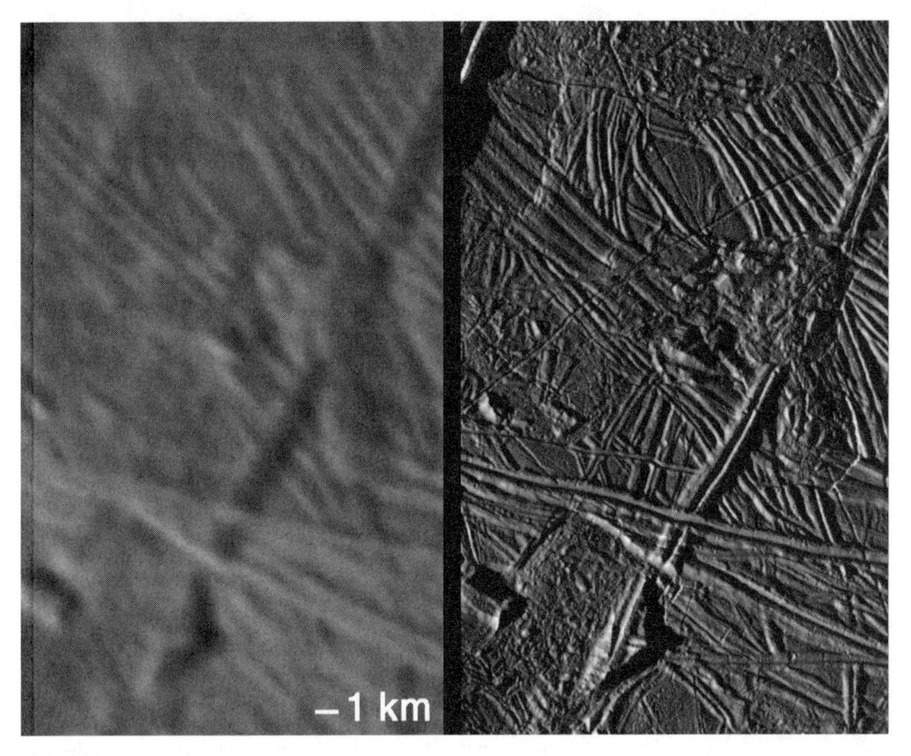

FIGURE 13.7: Small patches of chaotic terrain, at 230 m/pixel (left) and 34 m/pixel (right). Note that a couple of the rafts near the bottom are quite tilted.

comes into play in regard to one of the most enduring—and incorrect—arguments for thick ice on Europa, as will be discussed in **Chapter 14**.)

The high-resolution view in **Fig. 13.7** also shows a couple of good examples of one of my favorite phenomena, the tilted raft. To me these rafts look like foundering ships with one end sloping down into the water and the other end raised up. I suspect these things may be quite common, but they usually appear only under ideal lighting conditions, especially near the day-night line where shadows are long, as is the case in **Fig. 13.7**. In some images the long shadows exaggerate the tilts, making the rafts seem to stick up like shark fins. A systematic survey of these features is needed, since they might provide insight into the dynamics of chaos formation and evolution. How fast did the rafts move? How strong were the currents? Of course, any survey would have to account carefully for the lighting conditions, something that has been too often neglected in studies of Europa.

***○ *

Chaotic terrain looks like what sites on Earth look like where ice melts through from below, exposing liquid water, which then refreezes around blocks and rafts of drifted ice. When the Galileo imaging team saw the first detailed images of Conamara, this interpretation, sensibly enough, was prevalent. As plans were made for preparing a set of papers that would report the mission's preliminary results in the magazine *Nature,* imaging team member Mike Carr was put in charge of a paper that would follow this line of thinking, using the structure of Conamara as the then strongest evidence for an ocean on Europa. Carr was a senior planetary geologist from the US Geological Survey, independent of the Greeley and Head empires. He was already describing Conamara as a site of displaced rafts in a temporary melt-through zone well before I had given it much thought.

At that time, in early 1997, as the team's expert in tidal processes, I was deeply engaged in investigation of the tectonic record. While Conamara was interesting to me, and the melt-through explanation seemed obvious, chaotic terrain was tangential to the specific work I was doing. However, as a member of the imaging team I could hardly miss the fascinating human dynamics surrounding the preparation of the Carr *et al.* paper. Jim Head and his protègé Bob Pappalardo lobbied hard for their view that the ice was too thick to melt through and expose the liquid ocean. Ultimately the jointly authored paper had to accommodate both views, and not surprisingly it turned into a somewhat baffling compromise. Its title ("Evidence for a subsurface ocean on Europa") and much of its description of Conamara pointed toward oceanic exposure as the origin of chaos. But at the end the paper said that convection within the solid ice, independent of the ocean, was at least as plausible a cause.

Convection is most familiar to us in fluids like our atmosphere. On a hot afternoons, blobs or columns of warm air rise upward into the cooler high atmosphere, driving turbulence and often thunderstorms. Even solid ice can act like a fluid, if the forces acting on it are slow and continuous. Convection of a similar type is possible in an ice crust, but only if the ice meets certain criteria. First, the ice needs to be fluid enough, which depends on things like temperature and the size of the ice-crystal grains. The ice crust also needs to be thick enough for the vertical motion to get going. On Europa, this meant that the crust had to be thicker than 20 or 30 km, and even then it still had to have just the right fluid properties. So if you believed in thick ice, convection was plausible, although by no means a certainty or even a likelihood.

At the time of the publication of the Carr *et al.* paper, Pappalardo had only recently completed his graduate work, in which he focused on the geology of solid-state processes on other icy satellites. So it was natural for him to look at chaos on Europa from that perspective. He pointed to several examples of small round patches of chaotic terrain around Conamara (like the one in **Fig. 2.10**) as well as a variety of supposedly related features that he called "pits, spots, and domes." All them, he believed, were about the same size (~10 km), with fairly uniform spacing, and all of them, he said, represented the tops of convection cells, with their size and spacing implying an ice thickness of about 20 km.

In fact, it was arranged that Pappalardo would tell this story as first author of one of the team's articles in the issue of *Nature* that carried the article by Carr *et al.* And at about the same time, with his strong political backing, Pappalardo was designated as the imaging team's Europa spokesperson, assigned to present the team's early results and interpretations at various scientific conferences and policy forums.

The format of *Nature* is different from most scholarly publications. It is somewhere between a news magazine and a scientific journal. Most of the papers it publishes are in a very short format, quaintly called "letters". This format worked out perfectly for Pappalardo. He did not have to produce the supporting facts and details that would have been required for publication in most journals. Nevertheless, his *Nature* article seemed to have great credibility because it was an authorized presentation on behalf of the Galileo imaging team. So thick ice and the putative evidence for it became an accepted result of the Galileo mission. Never mind that it had not been quantitatively developed, and that its assumptions later proved incorrect.

If you did buy the idea that the 10-km-wide chaos patches were the tops of convection cells, then the model proposed that larger areas of chaos, like Conamara, were somehow the result of coalesced cells. How all that coalescing would have worked was always vague. Rising warm-ice blobs became a consistent theme of Jim Head and Bob Pappalardo's group at Brown University, including their explanation for double ridges (**Chapter 7**).

People tend to see on Europa whatever they are used to seeing. Ron Greeley has studied volcanism on the Earth and other planets. To him and his team at Arizona State University, chaos, especially the small round patches, looked like volcanic upwelling. Because cold stuff is involved, ice and water instead of rock and lava, this processes came to be called cryovolcanism.

A number of chaoses (as I like to call distinct patches of chaotic terrain) *do* look like upwelling volcanic flow. The little round chaos in **Fig. 2.10** is typical of this type. They appear to be bulged upward, their rounded edges are suggestive of viscous flow spreading over the adjacent crust, and sometimes cracks in the adjacent surface suggest that the weight of this "lava" broke the surrounding crust.

At first, I listened to the geological interpretations of the chaotic terrain with some detachment. I was working on the problems of tidal stress and tectonics. But the tectonic evidence was strongly arguing in favor of fairly thin ice. Our successful explanation of strike-slip displacement needed the cracks to penetrate to liquid, and the tidal stress seemed insufficient to crack down more than a few kilometers. Moreover, only cracks with double ridges were displaced by strike-slip, consistent with formation of ridges by tidal pumping of oceanic water to the surface. An integrated, coherent picture of thin ice—at least thin enough that cracks could link the ocean to the surface—was emerging from our studies of tidal tectonics. And thin ice was anathema to either the convection or cryovolcanism models.

While our thin-ice findings were perfectly consistent with the initial view of Carr (and nearly everyone else) that chaos represented oceanic exposure, thick ice was required for convection and for cryovolcanism. The convection model required ice thicker than ~20 km. And cryovolcanism would require deep pressurized chambers within the ice crust as the source for the upwelling flow of lava. The thin ice implied by our tidal tectonic studies was inconsistent with the emerging official stories explaining chaos.

So, as usual, I was in trouble. I had already been marginalized by the preliminary machinations within the imaging team that had divided up control of Europa science, and I was only allowed to calculate the effects of tidal stress because no one else could do it. Now I had results that contradicted the official thick-ice model. At this point my life would have been easy if I had jumped on the thick-ice bandwagon. But I am a scientist, not a politician. Having established several lines of evidence for thin ice, I set about trying to understand and evaluate the basis for the competing models.

The convection model was based on the putative "pits, spots and domes" that supposedly were the surface expression of convection cells. For four years nearly every presentation reviewing what Galileo had learned about Europa included a required mantra stating that Europa-is-covered-with-pits-spots-and-domes-that-all-have-about-the-same-size-and-spacing-and-demonstrated-that-there-was-solid-state-convection-and-ice-thicker-than-about-20-km. Yet the only documentation for that established fact was a couple of paragraphs in Pappalardo's *Nature* paper. Eventually, we showed that the taxonomy and generalizations about pits, spots, and domes were premature and not supported by a more complete survey (as described in **Chapter 14**).

The volcanism model seemed weak as well. Buoyancy presents a challenge for the notion of cryovolcanism because liquid water is more dense than ice. Ice in your drinking glass floats—it's on top of the liquid water. When an ice fisherman drills a hole through the ice, it does not open a gusher of cryovolcanism. The water stays below. In contrast, on terrestrial planets, volcanism occurs as the least dense materials, the magmas and lavas, rise to the surface. It is part of the global layering that puts the most dense materials in the center, making a nickel-iron core, and the least dense on the surface. On Europa, volcanic flow would require *ad hoc* pressure chambers to force the liquid upward, or additives to reduce the density of the water. In either case, the watery magma could not come from the global ocean: The ocean could not be pressurized enough to reach the surface, and if the density-reducing additives were widespread, the whole crust would have sunk into the ocean.

Despite their weaknesses, explanations that involved thick ice became the canonical models for chaos formation. To me, the evidence seemed superficial, preliminary, and highly questionable. The models certainly contradicted what we were learning from tidal tectonics. But solid-state upwellings and volcanism were the hobby horses of the team members who had been assigned control of Europa science, and that meant that thick ice was not going away.

The thin ice that we inferred from tidal tectonics fit much better with the original ideas about chaos, that it formed when the ocean melted through the ice from below. If the ice is thin enough, even modest local heating could cause occasional melt-through. This line of thought provided a simple explanation for why chaos looks the way it looks. And it fit with the evidence that the ice was thin enough to crack right through. I had not planned on getting involved in interpreting chaotic terrain. The ghost of Ockham made me do it.

I needed to roll up my sleeves and examine chaotic terrain more closely. Or, to be a little more accurate, I needed to roll up my students' sleeves and get them to examine chaotic terrain. Greg Hoppa and I spent a good part of the summer of 1998 looking over every available image in preparation for a systematic survey of chaoses. We pored over the Regional Mapping images, comparing them wherever possible with the few high-resolution images (as in the comparisons shown in **Figs. 13.6 and 13.7**). We even trained ourselves to recognize chaos when it appeared at or near the limits of recognizability (as in the left panel of **Fig. 13.7**).

At the end of the summer, undergraduate student Jeannie Riley was assigned by the NASA/Arizona Space Grant program to work with us. Jeannie was a smart and careful geology student. She quickly became skilled at recognizing chaos, and we set her to work mapping all the chaoses that appeared in the available data sets.

Before long we realized that patches of chaos show a continuous range of degradation, or changes in their features, brought on by subsequent tectonic processes, especially cracking and ridge-building. We had seen that Conamara, one of the freshest and most recent features on Europa, already

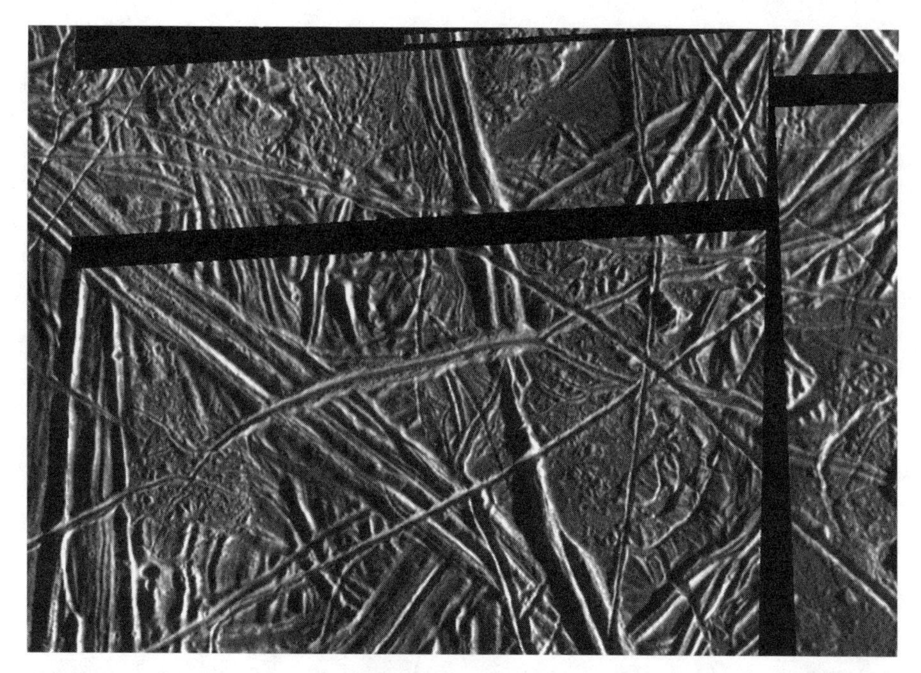

FIGURE 13.8: An area of complex interplay between chaos formation and tectonic resurfacing. It includes cracks and ridges crossing over older chaos, and disruption of tectonic features by chaos formation. This high-resolution (26 m/pixel) image is about 25 km across. Unfortunately, no images exist showing the regional context.

displayed signs of incipient degradation as cracks and ridges had already begun to cross over it. In other places, hints of older chaos were barely recognizable. **Fig. 13.8** shows an example of a locale where cracks and ridges have nearly obscured older chaotic terrain.

This realization added, to put it politely, a new dimension to our disagreement with the canonical description of Europa. The geologists had been creating maps of terrains and features. They used those maps to develop "stratigraphic" sequences that were supposed to represent long-term change in the character of the processes on Europa. These maps and sequences showed a lot of imagination. But there was one problem: The notion of stratigraphy as developed in terrestrial geology made no sense on Europa.

Terrestrial geological mapping largely depends on the idea that the history of the Earth's surface is recorded in a series of layers, or *strata*, that have been folded and crunched and tilted, then sliced off at the new surface. Thus a geological map shows the character and locations of the different layers where they intersect the surface, and it can be used to infer how the Earth changed over time. On Europa there are no strata, only a series of things that have modified the surface at different times and places. When traditional mapping protocols were followed without regard to the new context Europa presented, strange results were bound to come out. Geological mapping can be done meaningfully on other planets, but only by taking great care to review how the fundamental concepts fit the new environment.

The geological mappers for Galileo decided that chaos must be a relatively recent phenomenon, because the chaos that appeared on their maps was very fresh. They even found a theoretical explanation. A decade earlier, studying orbital change among the Galilean satellites, I had found that tidal heating might have decreased significantly over the past 100 million years, which would have allowed the ice on Europa to thicken relative to the past. The canonical model required thick ice, so by invoking my orbital-evolution theory, the geologists had a way to explain why chaotic terrain, as created by their proposed convection or cryovolcanism processes, only started to form relatively recently.

There was one strange upshot to this new coupling of one of my old ideas and the thick-ice theories. I was in effect being invited onto the bandwagon, and into a seat of honor at that. I certainly liked the idea that my earlier work on celestial mechanics might be supported by what we saw on Europa's surface. A theorist's greatest reward is when observations bear out predictions. But, as my group looked at the images, we saw no evidence that chaos was an especially recent phenomenon. The impression that it began to form only recently comes from the simple fact that the newer, fresher chaos was much more easy to spot than the older examples, which were obscured by subsequent processing of the surface.

When I presented this finding, Pappalardo seemed upset. He asked me how I could reject a result that was being touted as support for my earlier work on orbits. If I had been a good team player, I could have simply accepted the party line, and the canonical story would have incorporated my brilliant prediction as part of the obligatory mantra. In rejecting the scientific result about chaos, I was rejecting the favor as well. The problem was that as a scientist I felt the need to follow the evidence where it led, and that need as often as not is at odds with team spirit or even personal gratitude.

What's more, I came from a background in astronomy, where observational selection biases are an everyday issue. If a picture doesn't show asteroids that are small or dark, we don't take it as evidence that they don't exist. In fact, we usually assume that they do exist, and observers try to find ways to see them. Regarding Europa, the Galileo imaging team's canonical results were that most chaos is fresh and recent, and that most patches are larger than about 10 km across. But that geological party line ignored the observational selection effect created by both small and old chaoses: They are difficult to see and easy to overlook.

***O *

Jeannie Riley went to work and mapped chaotic terrain wherever she could find it. Remember, only about 10 percent of the surface was covered by Regional Mapping or by other images with 200m/pixel resolution. Also, we avoided images where the illumination was nearly vertical because that made surface forms and textures difficult to recognize. For example, the Brown University group had mapped the western Wedges region using images from orbit E14 with high sun angles, which contributed to their spurious result that most chaotic terrain is recent. Under those conditions, subtle older chaos would be especially hard to recognize.

In her survey, Jeannie took special care not to miss small or old examples of chaos. Among the very old examples, she paid special attention to a huge one (~1300 km across) near the center of the leading hemisphere. Although this giant chaos is relatively old, within it are patches of much fresher chaos that offer an important reminder: Formation of chaotic terrain disrupts whatever was there before. If the previous surface was tectonic, then cracks and ridges will appear on rafts. If the previous surface was chaotic, then pieces of the older chaotic terrain may appear on the new rafts. The continual resurfacing of Europa is by both tectonics and chaos formation. Tectonic processes break-up, cover, and destroy older terrain, and formation of chaos disrupts older terrain. In either case the older terrain might be chaotic or tectonic. In this way, Europa's surface has undergone the continual renewal that makes its surface so young compared with most other bodies in the solar system.

Then there was the matter of the small chaoses. Using image-processing software, Jeannie had marked all the chaos on the digital images, so it was straightforward to obtain statistical information. I was particularly curious about the statistics of the sizes of patches. The canonical model of convecting thick ice was founded on Pappalardo's claim that there was a characteristic, roughly 10-km, diameter for these features, corresponding to the scale of the supposed convection cells within the ice. As far as I could tell, the only supporting evidence for that notion was a qualitative impression from images of the region immediately around Conamara (**Fig. 2.7**) and the low-resolution freckles (the *lenticulae*) that I mentioned in Chapter 2. From Jeannie's maps, we had quantitative information that would allow us to test the claim that ~10 km was the norm.

We found that starting with the largest chaos, the number of patches increases rapidly with decreasing size. Roughly speaking, the number of patches of chaos measuring about 100 km across is only about 1 percent of the number of patches measuring about 10 km across. This increase in numbers as size gets smaller is so steep that there is about as much total surface area covered by 10-km-wide patches as by 100-km-wide ones. This steep curve does not, however, continue indefinitely. The statistics for Jeannie's maps based on the 200-m/pixel images showed a drop in numbers for patches smaller than 10 km. In other words, there is a distinct peak in numbers at ~10 km, seeming to confirm the underpinnings of the convecting-thick-ice model. This peak seemed to demonstrate quantitatively what Pappalardo and the Brown group had perceived. But true to my growing reputation as a spoiler, I had other thoughts.

I had worked with statistics like this before. The size distribution of known asteroids follows a similar curve, with a similar drop-off at small sizes. Astronomers understand that there is a limit to the sizes of bodies that they can see, so that drop-offs need to be looked at critically, since besides reflecting objective reality—what's actually out there in space—they just as well reflect our observational limitations. The peak in numbers of observed asteroids shows the size at which asteroids become too small to recognize. It is a measure of our observational limits.

How could we test whether the 10-km peak in chaos size was a similar observational artifact? We had already seen that chaoses that were marginally recognizable at 200 m/pixel were perfectly clear at 30 m/pixel. So we decided to use the high-resolution images to extend our survey down to much smaller sizes. Those images covered only a tiny portion of the surface, but we could scale the numbers to compare them with the results from the lower-resolution images.

These results show the effect of more complete recognizability at sizes below 10 km in the high-resolution data. Starting at the large sizes, the numbers increase with decreasing size. But rather than peaking at 10 km, the numbers continue to climb at smaller sizes. They only decrease for sizes smaller than a couple of kilometers. In other words, the most common size (where the number of patches of chaos peaked) was proportional to the resolution of the images. It was clear that the peak at 10 km was entirely an observational artifact. The characteristic size of patches of chaos depended on how the pictures were taken. The putative observational evidence for the convective thick-ice model was nonexistent.

As far as we can tell *from the observations*, the actual numbers continue to increase indefinitely with smaller sizes. We would quickly run into trouble with that extrapolation. There would be so many small chaoses between about 1 km and 10 km that the sum of their areas would be as great as the total surface area of Europa. The entire surface would be saturated with tiny patches of chaos, with no room for tectonic terrain. Clearly there are not that many tiny chaoses. However, we can get some sense of the statistics of the smallest patches of chaos from the highest-resolution images. If we add up all the chaoses from tiny sizes on up, we find that at least 30 percent of the surface, and more likely about 40 percent, is covered by chaotic terrain.

When the imaging team was initially focusing on the very fresh and obvious example Conamara, the impression was that chaos was recent and unusual. Conamara became the archetype for chaotic terrain. But our survey showed that chaotic terrain has been forming, as far as we can tell, throughout the geological history of the surface, and it covers nearly as much of the surface as tectonics, the other dominant type of terrain. Moreover, there is no evidence for a typical size of 10 km.

The foundation for the convective thick-ice model was shaky, and the tidal-tectonic work suggested that the ice was thin enough for cracks to penetrate to the ocean. The idea that chaos simply represented sites of melt-through of the ice crust seemed increasingly attractive. Either a portion of the surface ice completely melts or it becomes so thin that rafts can break free and move around. Either way, the ocean is exposed to space, but only briefly as, like a bird bath exposed to the night sky, the surface quickly turns back to ice. Such a process could create the observed characteristics of chaotic terrain.

It would be surprising if tidal heat from the deep interior of Europa came out uniformly around the globe. On any real planet—think of Earth, or especially Europa's neighbor Io—much of the heat comes out at particular, limited locations, where we see volcanoes. On Europa, too, we would expect local warm spots under the ice, so various parts of the crust could get a little extra heat at times. These local concentrations of heat might result from ocean currents, from tidal heating in the ice itself, or from hot spots in the rock below. But whatever the source of the heat on Europa, one can estimate its effect on the icy surface. Suppose a warm spot causes the ice to thin from below (**Fig. 13.9**). Because the ice is about 10 percent less dense than liquid water, it has a float line about 10 percent of the way from the surface to the base of the ice. That is the level to which water would rise through a fishing hole, or more likely a crack through the ice. (I am still amazed that I was able to sneak in the little ice-fishing house when I published this picture in *Icarus*.)

With more heat, as the ice thins further, the buoyancy decreases so the surface sags downward (**Fig. 13.9b**), keeping the float line level. If melting does not continue beyond this point, we might find a depression, which may explain the many "pits" seen on Europa. Eventually the ice thins so that

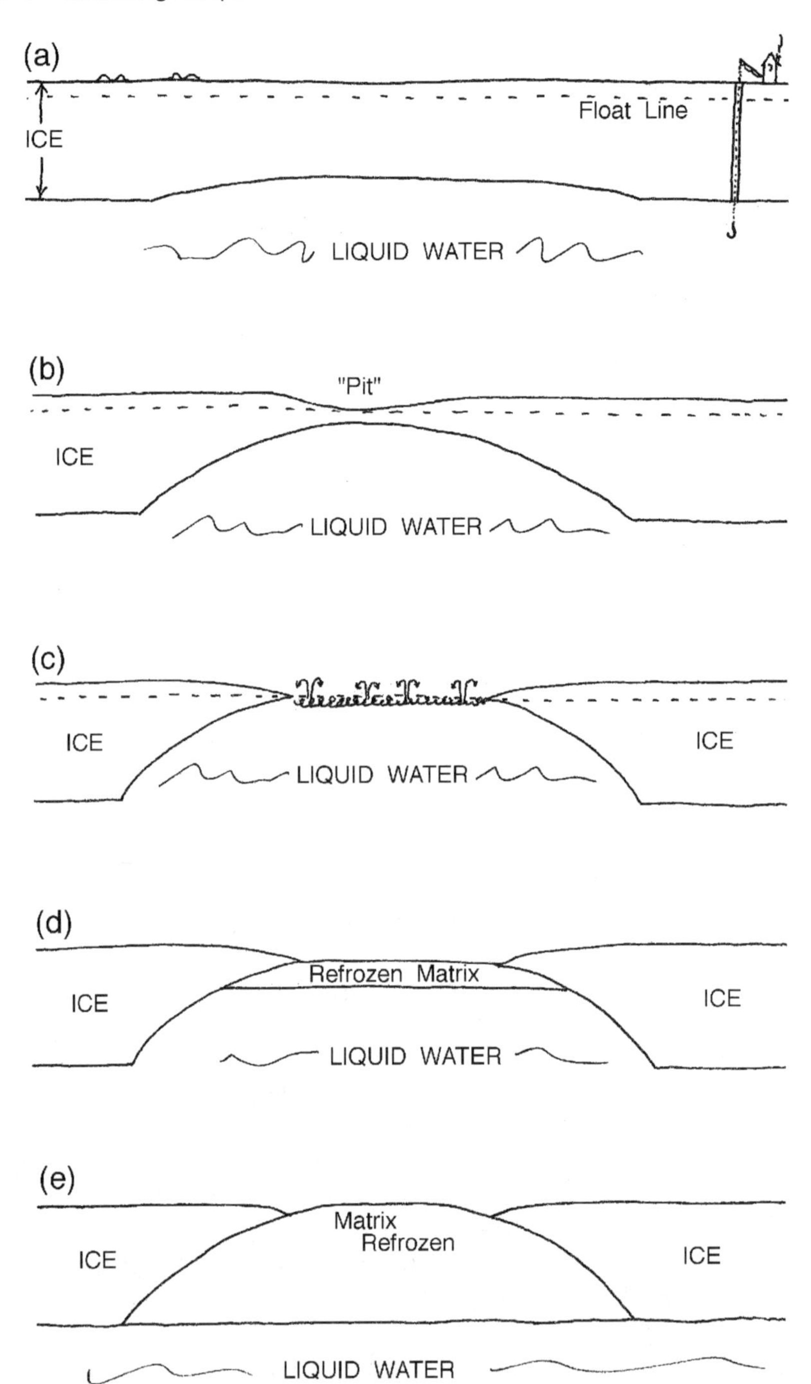

FIGURE 13.9: A schematic of the melt-through model of chaos formation.

liquid is either exposed at the surface or covered by a fragile, thin skim of ice (**Fig. 13.9c**). The surface tapers to the opening, with buoyancy ensuring that the edge of the ice is at the float line. The "shorelines" of chaotic terrain on Europa often display such ramped "beaches." Where liquid is exposed to space, its surface is agitated by boiling and freezing. The active surface might spray around material, including salts, organics, and impurities. The activity continues as long as adequate heat continues to warm the liquid. Then, as the heat source diminishes, the surface refreezes and a new crust forms and thickens (**Fig. 13.9d**).

The new surface would be rough and bumpy from the agitation of the boiling and from floating chunks of unmelted crust. Just as the numbers and sizes of rafts vary from case to case, the lumpiness varies as well, just as we saw in our survey of chaotic terrain.

As refreezing continues, the crust under the chaos area thickens until it becomes similar to the surrounding crust. Now the buoyancy of the refrozen ice raises the chaos area, with the edge forming an encircling moat (**Fig. 13.9e**). This process explains the common domed shape of many chaos areas rising from a gradually ramped shoreline (as visible, for example, in **Fig. 2.10**).

How do rafts form? In some cases the downwarping at the edge (**Fig. 13.9c**) might also result in cracking parallel to the shoreline. If this cracking penetrates through the crust (which is quite thin at this location), rafts may be launched and float away from the shore. Some may break up or partially melt. Flotsam in a wide range of shapes and sizes is likely to be created, ranging from small pieces that contribute to the texture of the matrix to large pieces that are part of the identifiable raft population.

Many chaos areas are bordered by cliffs, rather than by gradually sloping beaches, just as we would expect after shoreline rafts drift away. These two types of chaos appearance can be explained without any separate formation mechanisms. Ockham would love it.

In Conamara (**Figs. 13.1–13.5**) and in chaos in general, many rafts broke away from the edge, but many others represent the survival of pre-existing ridges (**Fig. 13.2b**). At a major ridge, the crust must be thicker than elsewhere, both by being higher at the top, and probably somewhat deeper at the bottom for support. As the crust is heated from below (**Fig. 13.9b**), the thicker ridge sites would resist melting, giving them (and the area immediately around them) a preference to survive as rafts.

The melt-through model also suggests that parts of the crust surrounding large chaos areas might have become quite thin during the process. It is no surprise then that large chaoses (like Conamara, for example) are often surrounded by numerous smaller ones; melt-through on a small scale is easier where the ice is thin.

The melt-through model also fits well with one of the most promising explanations for the darkening that surrounds most tiny patches of chaos, and also borders mature ridge complexes (the "triple bands"). Sarah Fagents, a post-doctoral researcher at Arizona State University, suggested that warming near the surface results in sublimation of the ice, concentrating the impurities and darkening the ice. In the case of ridge complexes, repeated passage of warm tidal water up through the crust would warm the nearby surface. Similarly, during chaos formation, warming of the adjacent terrain would have the same effect. In both cases, the terrain is not only darkened, but usually has topography that seems somewhat softened and subdued, consistent with Sarah's model.

An extreme example of this effect is the strange feature informally called the Dark Pool (**Fig. 13.10**). It is not a pool at all, but rather a small, flattened, darkened area of ice about 3 km wide . The Dark Pool lies in tectonic terrain, probably the fine furrows of an old dilation band just west of the area shown in **Fig. 13.8**, where considerable chaotic terrain has been formed during

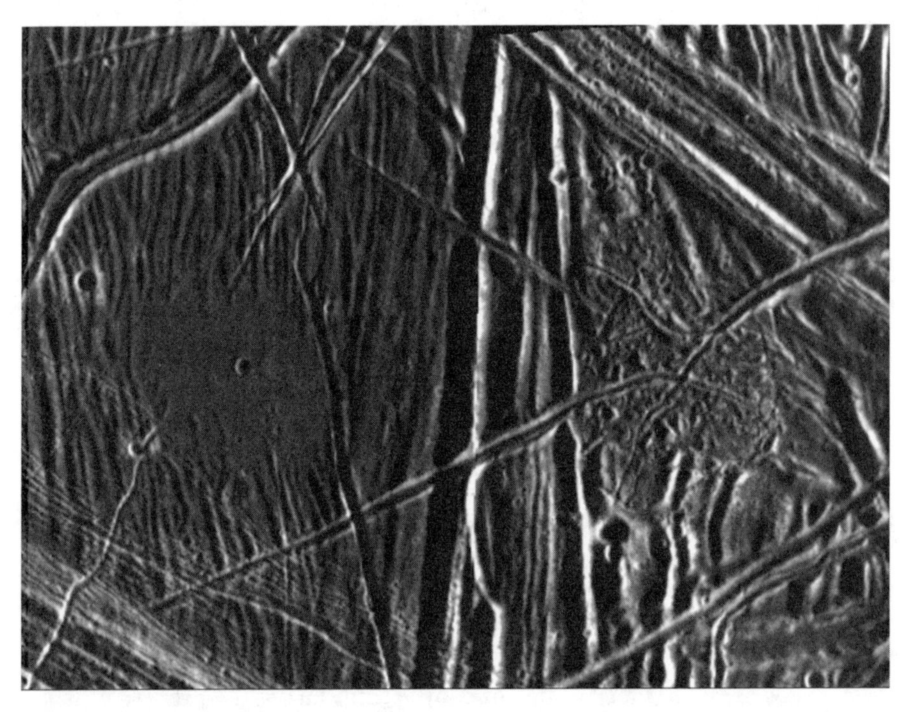

FIGURE 13.10: The Dark Pool is not a pool at all, but probably a place where warming from below has smoothed and darkened the surface. This image is about 13 km across, and unfortunately we have no context images to supplement this high-resolution view. The area is sprinkled with small craters, probably formed by material ejected from a larger crater nearby.

several episodes over time. In fact, included in **Fig. 13.10,** on the right, is a patch of chaotic terrain almost the same size as the Dark Pool. If Sarah Fagents is correct and heating from below causes the darkening of the surface, then the dark pool might be an incipient patch of chaos, where melt-through has nearly occurred (as in **Fig. 13.9b**). In that case, either the local heating under the ice diminished before complete melt-through could occur, leaving only a smoothed, darkened depression, or we are looking at a place where chaos formation is still in process.

I already described the main problem with volcanism: Buoyancy tends to keep denser liquid (water) down under the less dense surface (floating ice). I suppose you could call our mechanism for producing ridges a kind of volcanism, as diurnal tides squeeze water and slush up and onto the surface along the sides of the cracks (**Fig. 7.2**). But it is difficult to envision a theoretical mechanism that could create chaotic terrain with volcanism, unless *ad hoc* conditions are invented.

Some examples of chaotic terrain do look a bit like upwelling lava, especially the small rounded ones (**Fig. 13.11**). The case for such upwelling is based on the topography. The lumpy matrix bulges upward, and you could imagine that it has flowed over the surrounding terrain. Indeed, this very

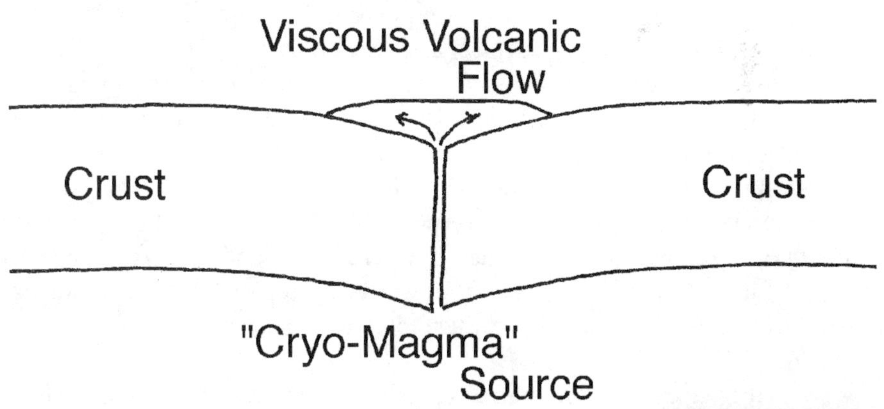

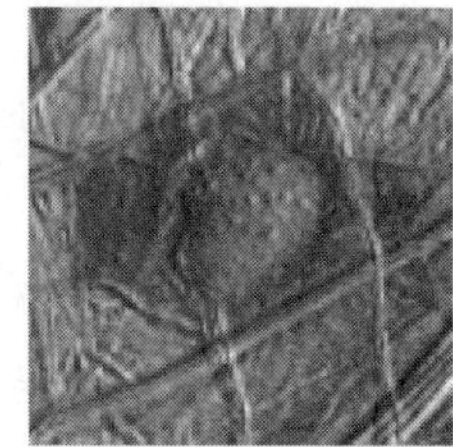

FIGURE 13.11: Volcanic flow of a viscous fluid up and over the crust (above) might explain the apparent topography of many small round chaoses, such as the example at the right (already seen in **Fig. 2.10**). However, exactly the same topography can be created in the melt-through model—in which the features result from the buoyant relaxation of the crust after refreezing, at the end of the process shown in **Fig. 13.9**.

scenario would be a way to explain the slope of the surrounding terrain, toward the edge of the matrix, where the crust might be bent down by the load (**Fig. 13.11**). No wonder that such patches of chaotic terrain would give the impression of volcanism to someone in that frame of mind, especially a volcanologist like Ron Greeley. The problem is that it only applies to certain patches of chaos, those that have this particular appearance.

On the other hand, these features are equally well explained by the melt-through model. The topography created by melt-through followed by refreezing would be bulged up so that it might look like a viscous lava flow. (In fact, the surface profile at the last stage in **Fig 13.9** is identical to that in **Fig. 13.11**.) From a philosophical point of view, the melt-through model is also supported by Ockham's Razor, because it explains not only the bulged up appearance of some examples of chaos, but also all the other characteristics of chaotic terrain in general.

One potential way to discriminate between the volcanic cross-section and the refrozen melt-through cross-section would be to measure the surface topography. If in general the updomed matrices of chaoses proved to be higher than the surrounding terrain, it would suggest that some active upwelling, not just buoyancy, was at work. So far, however, there is no good evidence that chaos terrains, as a rule, bulged upward that far. But there is bogus evidence for this phenomenon, and we will return to discuss it in **Chapter 16.**

Even as I began to make the case for melt-through, the heavy hitters on the Galileo imaging team were widely reporting the discovery of numerous volcanic flows over Europa's surface, and there was considerable attention and image-sequence planning devoted to getting high-resolution images of them. The volcanic flows Greeley and Head thought they were seeing, based on low-resolution images, found no believers within my research group, however. Greg Hoppa and Randy Tufts in particular thought the talk of volcanic flows had no basis in the evidence. Whenever we had looked at Europa's dark splotches closely (that is, wherever we had high-res images), the splotches had turned out to be chaotic terrain. Applying Ockham's Razor again, we expected that the other dark splotches would, similarly, turn out to be chaos.

And at better resolution that is exactly what they proved to be. For example, the areas known as Thrace and Thera had been touted as likely lava flows. These two large, dark regions had first been identified southeast of the Wedges region (they are labeled in **Fig. 6.1**) in low-resolution images from the Voyager spacecraft. Then they were imaged in color during Galileo's orbit E17 (**Color Plate 5**). As we expected, Thrace and Thera are not lava flows, but simply large, classic examples of chaotic terrain.

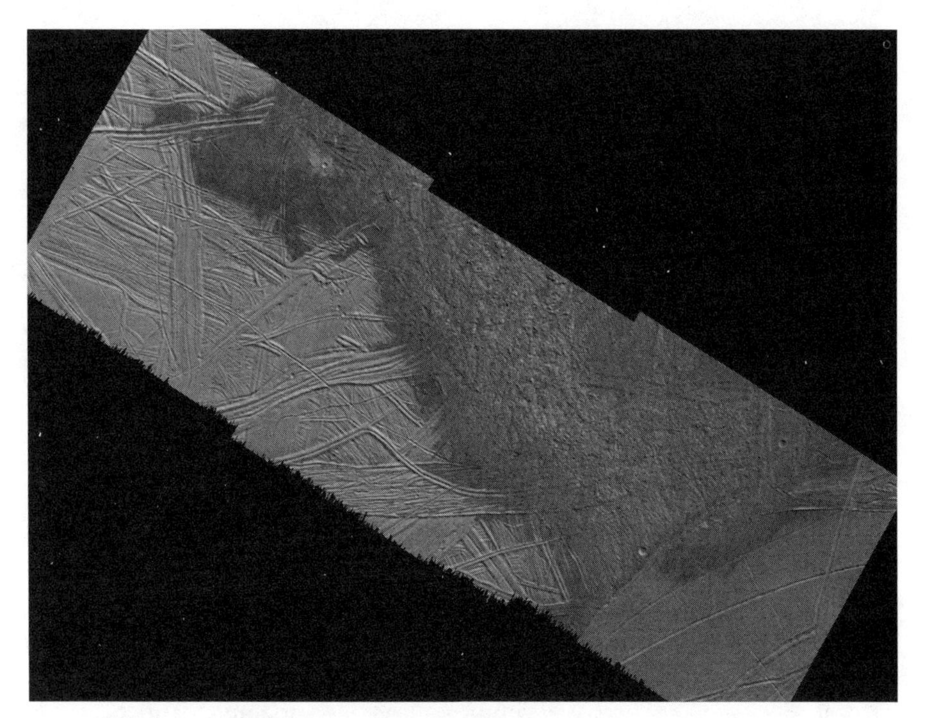

FIGURE 13.12: Thrace's southwest margins.

During the same orbit, high-resolution images were taken of the southwestern margins of Thrace (**Fig. 13.12**) because Jim Head wanted to get a good look at his lava flows. And that is exactly what he saw: dark margins around the chaos area that he interpreted as flows of darkened material that seem to have filled in between higher ridges.

Ockham's Razor insisted on a different interpretation. Similar darkened terrain is quite common, as we saw in **Chapter 2**: the margins of mature ridge complexes, which appear in low resolution as the dark components of the low-resolution triple bands; and the dark surroundings of chaotic terrain, which for the smallest chaoses appear at low resolution as the dark spots named lenticulae. Wherever we find these darkened terrains, they presented similar characteristics: Slight softening of the topography, and darkening in the lower parts of the flattened portions. This common appearance in those locations is consistent with Sarah Fagents' model, in which heating has both softened the surface and darkened it where the most sublimation occurred.

The softened, darkened appearance at the edge of Thrace fits the same pattern seen along the margins of cracks and other chaos. And those locations are precisely where we would expect warming to be present, whether because of the tidal working and oceanic flow through cracks, or because of the melting that leads to chaos. There was no need to invoke an entirely different explanation, volcanic flows, for this location near Thrace. It is a natural consequence of the melt-through process of chaos formation.

The melt-through idea is attractive because it explains all the properties of chaos with a single process. And that process follows naturally from the tidal-tectonic evidence that the ice is relatively thin, thin enough that the ocean has been linked to the surface through cracking and melting. Ockham's Razor implores us to favor simple explanations that cover multiple phenomena, and that logic compelled us toward the melt-through model for chaos. Outside my own research group, however, the imaging team's geologists were not so compelled by the ghost of Ockham. They had become too deeply invested in the idea of thick ice.

***O *

The advocates for the thick-ice model fought vigorously against the idea of melt-through, using their political advantage to promote the canonical model. It was arranged that Pappalardo would put together a paper for the *Journal of Geophysical Research* defining the Galileo imaging team's position on Europa's geology in the context of the key issue of the time: "Does Europa have a subsurface ocean: Evaluation of the geological evidence." The subtitle suggested a balanced, authoritative assessment. Unfortunately, the paper was an advocacy piece for thick ice, concluding under the mantle of authority of the imaging team that there was no evidence for any direct linkage between a liquid water ocean and the surface. It became the centerpiece of the isolated ocean model.

To show that the melt-though model was physically impossible, they used the following argument. (I am not making this up.) Even if you took all the internal tidal heating from all of Europa and concentrated it all right under Conamara, and even if the crust were only 2 km thick, it would take hundreds of years worth of heat to melt the ice, they said. How this showed that melt-through was impossible is beyond me. I guess hundreds of years seemed like a really long time. But why should that length of time be a standard for testing the plausibility of our model? Who's to say that Conamara did not take longer to form?

A relevant time scale for comparison would be the rates of other geological changes on Europa. We have estimated that ridges form in tens of thousands of years (in **Chapter 7**), and that diurnal tidal stress undergoes changes over a similar length of time due to nonsynchronous rotation (in **Chapter 5**). So chaotic terrain formation could take that much time and still not be interrupted much by tectonic processes. We can repeat the calculation by Pappalardo *et al.* using this amount of time instead of their totally arbitrary time span of hundreds of years, but otherwise making the same assumptions, including the same tidal heating rate. It turns out that only 1 percent of Europa's tidal heat budget would be enough to melt through all of Conamara in about 10,000 years. There is no problem producing a large area of chaos like Conamara in a geologically short period of time. And it is even easier with the much higher tidal heating rate estimated from a comparison with Io's measured tidal heating (**Chapter 5**). The rhetorical trick of using an arbitrary criterion (require formation within hundreds of years) in the *JGR* paper could make melt-through sound silly to an uncritical reader, but appropriate arithmetic shows it to be completely reasonable.

Another objection to the melt-through model was more thoughtful, although ultimately flawed. In 2000 David Stevenson, a planetary geophysicist at Caltech, noted that the warm ice at the bottom of the crust, just above the liquid water ocean, would flow and tend to level off the bottom of the ice. For example, if we consider **Fig. 13.9b**, the warm ice from the sides would tend to flow up and into the thin spot. Just as a fluid on the surface of a planet tends to flow down into a depression due to gravity, the viscous ice on the bottom of the crust would flow up into the thin zone due to buoyancy. On the basis of this idea, Stevenson declared that melt-through was impossible.

But such pronouncements proved to be premature. My graduate student Dave O'Brien, who had been working on asteroids, naturally attended the weekly meetings in my lab, and followed the issues raised by our study of Europa. Using computer simulations, he confirmed that the ice can be melted through in reasonable time with only a very modest concentration of the tidal heat.

The computer-generated cross-sections of the crust during the melting process looked uncannily like my earlier sketches of the melt-through process (**Fig. 13.9**), but without the fishing shack. Moreover, Dave O'Brien's simulations showed that the bottom of the ice melts away so fast that inflow from the edges of the melt-zone envisioned by David Stevenson is not a factor. Stevenson had raised an interesting issue, but Dave O'Brien had shown that it was not an impediment to melting holes in the crust.

Dave O'Brien presented his results at the autumn 2000 meeting of the DPS (the Division of Planetary Sciences of the American Astronomical Society). DPS presentations are limited to a few minutes, and the open discussion after each is even shorter. Questioning can be brutal, but usually student presenters are treated relatively gently. This time, though, was an exception: Stevenson proclaimed before the large audience that O'Brien's work was all wrong. Of course there was not enough time for any explanation of what he meant by that, but the damage was done. The audience had heard the definitive judgment by a renowned authority that this graduate student's work (which bucked the Galileo party line to boot) was worthless. Later, during a private conversation, Stevenson agreed that Dave O'Brien's work was fine. But the DPS audience never heard that, and the damage was done.

Formation of chaotic terrain requires modest concentration of heat from below at various times and places. Are such concentrations plausible? On Io, a whopping 20 percent of the total tidal heat escapes through one volcano—a single, solitary volcano! On Europa, the modest, occasional concentrations of heat adequate to melt through the ice crust might also come from volcanic vents. Undersea volcanism on Europa has been proposed as a way to support ocean life, and is widely considered plausible. Astrobiological theorists were constrained by the canonical wisdom about Europa, which they had received through the authorized Galileo channels, that the ocean was isolated from the surface. They thought they needed to invoke volcanism as a source of the energy and chemistry for life. More likely, the ocean is linked to the surface, so the possibility of life is much less constrained, but the plausible volcanoes, though not essential to making an argument for life below the surface ice, could nonetheless play a role in localizing the heat output from Europa's deep interior. (More about the relevance of sub-oceanic volcanic vents to Europa is discussed in **Chapter 16.**)

As Galileo imaging was providing evidence for an ocean, Galileo's Project Scientist Torrence Johnson encouraged John Delaney, a prominent geophysicist from the University of Washington, to consider the role of volcanism in supporting oceanic life. Delaney had led important explorations of life at sub-oceanic hot springs on Earth. Delaney in turn recruited Richard Thomson, an oceanographic hydrodynamicist from the University of Victoria in British Columbia.

Thomson and Delaney examined the critical issue of whether the concentration of heat at a volcanic vent under the ocean could remain concentrated as it moved outward through 100 km or more of liquid water. As a Canadian oceanographer (thus not beholden to the US government agency NASA) and a researcher of integrity, Thomson had no particular interest in supporting the Galileo party line, nor did Delaney. They concluded that thermal plumes rising from undersea hot-spots could support our picture of melt-through as the origin of chaotic terrain. At least within our range of uncertainty about the character of the ocean, melt-through remains a perfectly plausible process.

Study of chaotic terrain implies that melt-through occurs at various times and places. It follows that the ice thickness varies with time and place. Chaos formation has been common, currently covering nearly half the surface, and continual, displaying a wide range of degrees of degradation by subsequent cracking and ridge formation. The sizes of chaos patches range from over 1000 km across down to as least as small as 3 km across—that is, what's recognizable on available images. Early reports that chaos is rare and recent, or that there is a characteristic size to the small patches, were observational artifacts: Old or small examples are harder to see, but they are by far the most common. Over the surface age, formation of areas of chaotic terrain has been frequent and continually interleaved with tectonics, so at any given location melt-through has likely occurred roughly every few million years. We will come back to that number in **Chapter 17**, where we consider the prospects for life on Europa.

Thick vs. Thin

One need not on that account take the common popular assent as an argument for the truth of what is stated; for if we should examine these very men concerning their reasons for what they believe, and on the other hand listen to the experiences and proofs which induce a few others to believe the contrary, we should find the latter to be persuaded by very sound arguments, and the former by simple appearances and vain or ridiculous impressions.

—*Galileo, 1615*

A 1999 article in *Science* first got me excited about the new science of "astrobiology," although it probably had the opposite effect on most readers. The article—by a young post-doc at Caltech, Eric Gaidos, and a couple of his senior colleagues, Ken Nealson and Joe Kirschvink—seemed to put a damper on the rising speculation about life on Europa. Gaidos et al. considered the implications of the canonical interpretation: that the ice crust on Europa was so thick that any ocean must be isolated from the surface. Having accepted the prevailing thick-ice dogma, they then concluded that, because the ocean was separated from chemical oxidants at the surface, there must be severe limitations on life on Europa. If life existed at all, it would be minimal in its isolated and sealed liquid ocean.

Even with speculation about chemistry and energy from possible undersea volcanism, life would be limited. Giant squid with "eyes the size of dinner plates" (as proposed tongue-in-cheek by Chris Chyba of the SETI Institute) were ruled out as were the weird advanced forms of Europan aquatic life appearing as fanciful illustrations in popular science magazines. All these speculative forms of life were snuffed under the official thick layer of solid ice.

Of course I read the article differently.

My research group had already found strong evidence that the ocean was linked to the surface by cracking and melting of the icy crust. From that perspective, the implication of the Gaidos et al. paper was not that life was limited, but that the frequent and widely distributed openings through the ice could play a crucial role in supporting significant life on Europa. Our work took on a new importance, and at just the point at which NASA was promoting astrobiology as a major new initiative.

I met Gaidos for the first time just before a special session on astrobiology at an AGU (American Geophysical Union) meeting in San Francisco, where I had been invited to give a presentation. I told him how much I had enjoyed his article, because it had highlighted the importance of the linkages between the ocean and the surface. He looked at me as if I were insane: How could there be any linkage? Then he recited the known facts about Europa: The "pit, spots and domes" proved that the ice must be at least 20 km thick.

I could hardly blame the guy for believing it. Nearly every post-Galileo presentation about Europa had included among the basic known facts that "the surface is covered with pits, spots and domes, known as lenticulae, that are rounded, often up-domed and cracked across the top, regularly spaced, and typically about 10 km across, demonstrating solid-state convection in ice, which is thus at least 20 km thick".

Over five years I must have heard this identical mantra dozens of times, to the point that I could silently move my lips along with the speakers' words. These "facts" were codified in official reports of NASA and the National Research Council of the National Academy of Science. They became the basis for future mission planning and for decisions on research funding. Project spokesmen ensured that the canonical facts about Europa were duly reported in the media.

Naturally, when Gaidos did his background research, to learn what was known about Europa before making his own contributions, he had absorbed the party line. So he knew about the "pits, spots and domes" and what they implied. It was obvious that I was nuts to consider that the ice was thin enough for ocean-surface linkages. He had no reason to delve into the original sources or to pore over all the Galileo images, and he never suspected that the canonical facts were wrong.

The original source in the literature was the two paragraphs in Pappalardo et al.'s letter to *Nature* that I described in the last chapter. It included postage-stamp cut-outs of putative examples of "pits, spots, and domes." Readers usually assume that substantial evidence lies behind the claims in such short papers, but that trust can be abused. With the evidence of the pictures, the full authority of the Galileo imaging team implied by the publication format, the prestige of publication in *Nature*, and constant reinforcement by the repetition of the thick-ice mantra, who could question the facts?

***O *

At about the same time as Pappalardo et al. published the short *Nature* piece on PSDs ("pits, spots, and domes"), we published our first paper on tidal tectonics. We realized that its implications for permeable ice—ice thin enough for cracks to get through—was contrary to what the powers in the imaging team were broadcasting so effectively. We certainly understood the risk in contradicting the party line. On the other hand, as Randy pointed out, there was also an odd kind of advantage: Our ideas were so obviously contradictory to the accepted facts, there was very little likelihood they would be appropriated by others.

That issue is significant because it is easy to get ripped off in the ruthlessly competitive world of Big Science. Here is how things worked in the Europa project. If you had a good idea and tried to be a team player, the idea would be subsumed into the team's accomplishments. Then the powerful players assigned to present the results would take the credit. Remember my ideas about digital image processing: first mocked, then implemented, with Ron Greeley picking up the marbles. Another example comes from the early 1990s, when the Galileo camera discovered a tiny moon orbiting asteroid Ida. I had assembled evidence from the moon's orbit that showed Ida had a very low density, and we discussed it during a meeting of my research group. Mike Belton, the Galileo Imaging team leader, happened to be visiting our building and wandered in, so I proudly shared our result. My students were amazed when Belton authored the paper reporting this discovery.

So, with Greeley and Head designated as the imaging team's leaders for Europa science, we worried that our findings would again be subsumed into the team's results, with you-know-who taking all the credit. From that point of view, it was actually good that the implications of our work were so clearly different from the developing isolated-ocean party line. In fact, we were so far out of line politically that we were totally free to follow the evidence where it led us. Generally it is no fun being a pariah, but at least we had that compensation.

We always tried to characterize the difference between our view and the mainstream view not as thin ice vs. thick ice, but as our notion of permeable ice vs. the canonical isolated ocean. That qualitative distinction is much more significant than any specific numbers. The isolated ocean model depended on convection, and that needed ice thicker than 20 km. To be permeable, the ice must be considerably thinner. Just how much thinner I do not know (although I have often been misquoted as advocating a specific value). But a difference between numbers is not the issue. What really set our results apart were the direct and active connections between the ocean and the surface. That is the story Europa's cracked and melted and shifted surface told us.

Because the case for thick ice had such powerful political support, we could not simply provide evidence for permeable ice. We had to explain what was wrong with the case for the thick-ice, isolated ocean model.

***O *

"Pits, spots, and domes" have never been well defined as a class or as three separate sub-classes. Beginning with the *Nature* letter in 1998, they have only been represented by examples (see **Fig. 14.1**), along with unsupported claims that they have certain general characteristics. Supposedly, PSDs are all:

- about the same size, 10 km across;
- regularly spaced about 20-km apart;
- roundish;
- and their domes are often cracked on top.

If you believe it, those characteristics add up to convincing evidence for solid-state upwelling and convection. And what was not to believe? The *Nature* piece showed them all!

The problem is, the story unravels when you take a careful look. Consider the claim about equal sizes. The six examples shown in the *Nature* report are

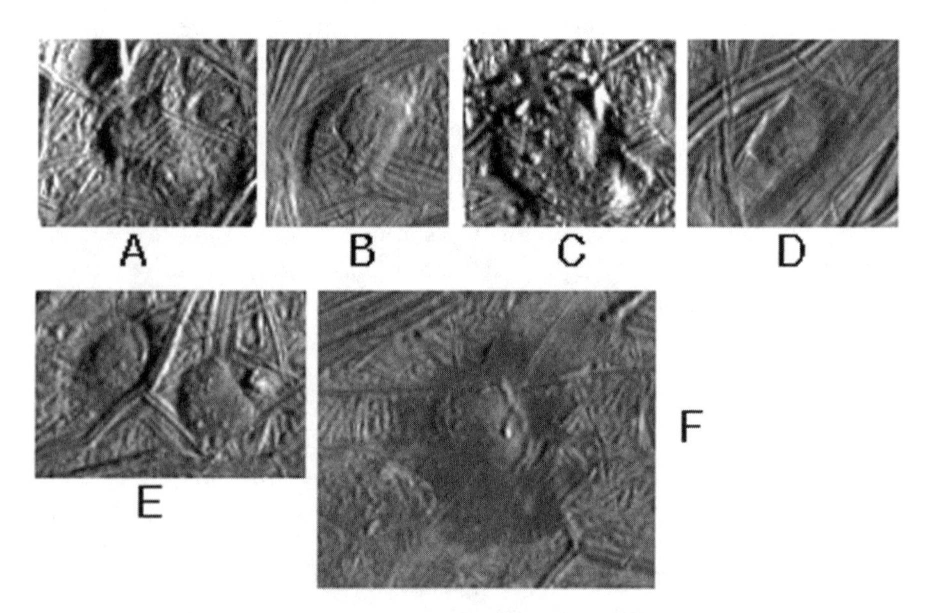

FIGURE 14.1: The six supposedly defining examples of PSDs, exactly as presented in Pappalardo *et al.*'s 1998 *Nature* paper. These pictures were snipped out of images of the area around Conamara, which we saw in **Fig. 2.7**. The six segments are shown here all at the same scale, with the height of frames A-E about 15 km. A and B are irregularly shaped uplifts, contrary to the *Nature* paper's statement (made repeatedly) that all PSDs are "circular to elliptical." The other four segments (C, D, E, and F) are simply examples of chaotic terrain of a *selected* size. Contrary to the impression given by this selection, patches of chaotic terrain are found in all sizes and shapes, and not with the "typical" size and shape implied here.

postage-stamp cut-outs from broader views. They nearly all come from one Europa neighborhood, the area around Conamara, as it appears in this book in **Fig. 2.7**. In fact, most of the examples are small patches of chaotic terrain. Chaos patches, as discussed in **Chapter 13**, come in all sizes, so the selection in the *Nature* article was misleading. What's more, one of the six examples (B in **Fig. 14.1**) is an uplifted feature that has nothing in common with chaotic terrain. Apparently, its sole qualification for being included with the other selected examples was that it matched them in size.

Suppose an alien biologist had visited Earth and returned home with an official report that life here is dominated by animals typically 1 meter tall. The mission report includes half a dozen photos of six-year-old kids and a pony, all 1 m tall. For five years every scientist describing Earth includes the basic revealed knowledge that animals are typically 1 m tall. The result is dogma, but is it true? A look at the original photos shows that each kid's picture was cut from a larger picture that included their parents and (harder to spot) tiny babies. People are not typically 1 m tall. And what's with that horse?

I felt that way about the PSD sample set that appeared in *Nature*. Like that pony, some of the PSDs presented as archetypes did not fit, independent of the size issue. Domes were supposed to be round, but the archetypical uplift feature in image B of **Fig. 14.1** was polygonal. In fact, most uplifts on Europa are fairly irregular in shape—not "roundish." And while that example is cracked, hardly any others are. The verbal description of round, raised, cracked domes fit the convection model perfectly. It just didn't describe what is on Europa.

Another thing that bothered me was a rhetorical trick that was part of the PSD mantra. Head, Pappalardo, and their associates repeatedly used the word *lenticulae* as a synonym for their PSDs, as in "Small pits, domes, and dark spots, collectively 'lenticulae,' pepper Europa's surface." But remember, *lenticulae* were clearly defined very differently by the IAU as the small dark spots that appear in certain low-resolution images on which the definition was based. As I discussed in **Chapters 2 and 13**, most *lenticulae* turned out to be dark haloes around very small patches of chaos, and there is a physical reason that those faint dark halos tend toward roundness with a typical size of ~10 km. By misusing the word *lenticulae*, and thus giving the impression that *lenticulae* were identical to the ill-defined set of things that were called "pits, spots and domes," Head, Pappalardo, *et al.* could use the legitimate properties of lenticulae (their sizes and shapes) to reinforced the false generalizations about the PSDs. It was a very effective sleight-of-hand.

It was really eye-opening to me to see how terminology could be used and misused in a descriptive branch of science like geology. As Bill Bryson has written, "Taxonomy is described sometimes as a science and sometimes as an art, but really it's a battleground." I wish I had realized it at the time.

The only thing that the features that were used to define PSDs have in common is their size. They were selected that way. In later publications, the authors even admit it. So the discovery that PSDs are all about the same size is not a big surprise (just as it should be no surprise that all one-meter-tall people and animals are the same height). And this characteristic size should certainly not be taken as evidence for a common formation mechanism. Yet it has been repeatedly cited (starting with Pappalardo's piece in *Nature*) as the factual observational underpinning of the argument for solid-state convection, and hence thick ice and an isolated ocean.

None of the claimed facts about PSDs are true. The characteristic size is a product of artificial selection; the characteristic shape never existed, even in the defining sample; the uniform spacing was never demonstrated and gradually faded from the story; and it turned out that the cracked tops are rare.

For all these reasons, the "Pits, Spots, and Domes" mantra rang false. Most of the examples were chaotic terrain, and our survey showed that chaos in general fit *none* of the supposed characteristics of PSDs. However, I was intrigued by the interesting pits on Europa (**Fig. 14.2**) and their possible connection to chaos (see **Fig. 13.8b**). Weirdly, of all the examples of features that were used to define "Pits, Spots and Domes" (eventually about 16 different examples were shown), *none* were pits. Go figure. And having noticed that none of the "Pits, Spots, and Domes" were pits, I was not surprised that none of them are domes, either. We saw plenty of uplift features on Europa, but few of them fit the description of "domes." Like chaotic terrain, uplifts are seen in a wide range of shapes and sizes, and usually look like pushed-up plateaus where the surface has been minimally disrupted (if at all) by the rising. Uplifts that are 10 km wide, round, and cracked on the top (as PSDs are supposed to be) are the exception, not the rule.

So our preliminary look at pits and uplifts made us even more skeptical about the claims regarding "pits, spots and domes." Key components of the case for thick ice were clearly wrong. Jeannie's survey of chaos was crucial because most PSDs were patches of chaos. And her results showed that the essential claim of a characteristic 10-km size was nonsense. It was aggravating to listen to the pit-spots-domes mantra, but we needed a comprehensive survey of pits and uplifts, in addition to our survey of chaos, to thoroughly refute it. So we put just such a survey on our list of things to do.

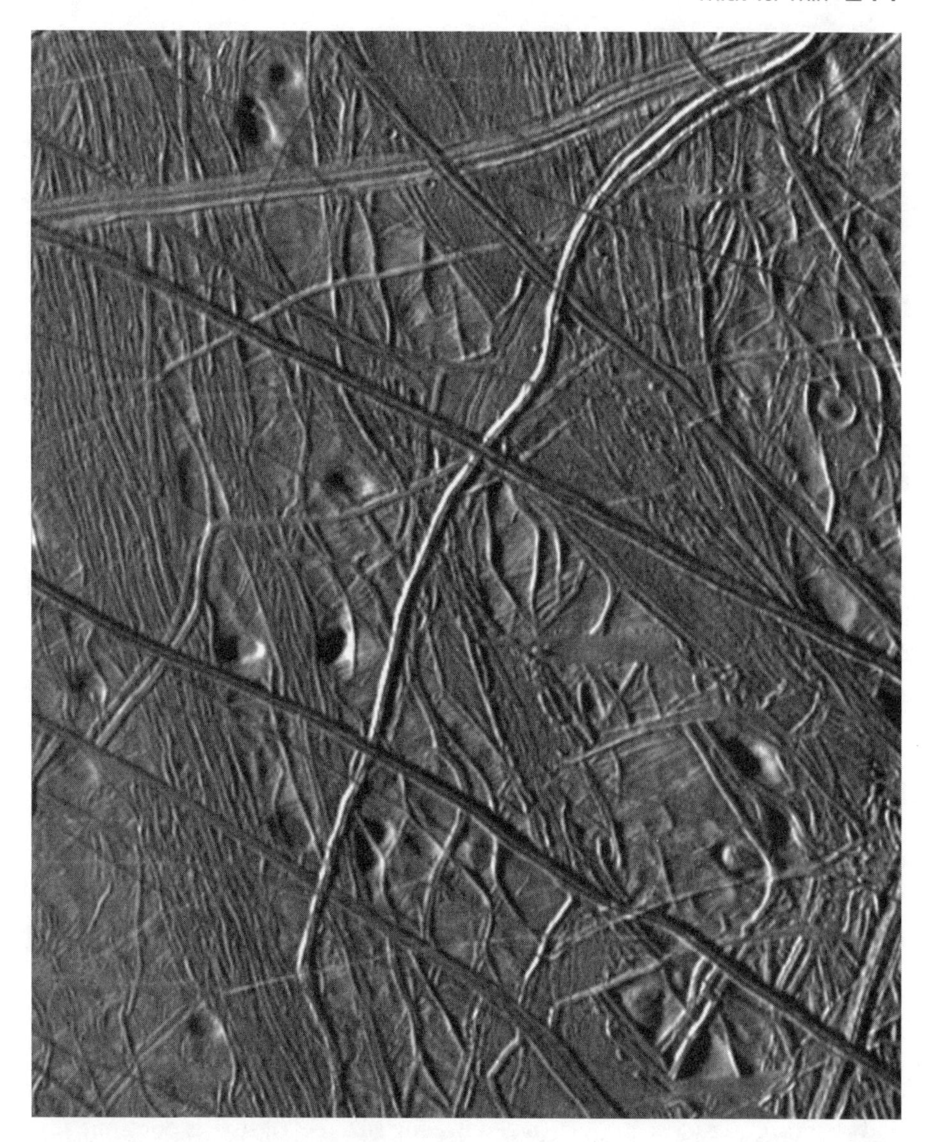

FIGURE 14.2: Examples of pits near 45°S,90°W. The area shown is about 100 km across at 200 m/pixel. Illumination is from the left (west). If your brain interprets this picture as if the light were from the right, the pits may seem like bumps to you. It might help to stand on your head or turn the book upside down.

Then, in late 2000, Martha Leake, a former Arizona student, came to work with us during a sabbatical leave from Valdosta State University, in Georgia. As a professor, Martha did a great deal of teaching, but the sabbatical gave her a chance to apply her expertise in planetary geology. She joined us in carrying out a complete and systematic survey of all the pits and uplifts in the Regional Mapping images, covering the same areas that had been surveyed by Jeannie Riley for chaotic terrain and by Alyssa Sarid for strike-slip.

The pits are a real and distinct class of feature. They were noted very early in the Galileo mission in low-resolution images. Taken at first to be small craters, they implied a fairly old surface. However, that age conflicted with the paucity of larger craters, which implied a younger surface. Gene Shoemaker, a grand sage of impact-crater studies, suggested one way to reconcile the statistics: The surface is very old, but larger craters forms have sagged so much that they cannot be seen. It was a nice try, but wrong. A closer look showed that the pits nestle between ridges, which would be hard to explain if they were impact features. Moreover, higher-resolution images later showed they are not shaped like craters. As discussed in **Chapter 13**, the pits may be the result of heating that thinned the ice, in sort of an aborted melt-through (as shown in **Fig. 13.9b**). That would explain why pits aren't found on ridges, where the crust by definition is thicker.

Martha Leake's survey allowed us to identify general properties without depending on anecdotal evidence. Pits differ from chaotic terrain in that there is no evidence of disruption of previous surface, and they lack the distinctive lumpy texture. The adjacent terrain, especially fine tectonic structures, seems to continue across the pits uninterrupted except by the change in topography that the pits themselves produce. In a few rare cases (<1 percent) the bottom of the pit is dark. In some cases the bottom of the pit seems smoother than the surrounding terrain, but in most cases it is hard to tell, given the small area involved and the shadows cast by the side of the pit.

These pits really are depressions, and in that sense they differ from all the PSDs. Shadow measurements and stereography indicate depths of 200-300 m for pits ~10 km across; smaller pits are not measurable but appear proportionately shallower

Uplifts tend to be irregular in shape, often polygonal and rarely rounded, and also rarely cracked on top. In general the surfaces of uplifts continue the character of the surrounding surface, as if the crust had been simply punched up from below. A large uplift, about 40 km long, is shown in **Fig. 14.3**. It seems fairly flat on top, is of an irregular shape, and has steep edges dropping down to the surrounding terrain, several hundred meters below. The terrain around this feature includes typical chaotic terrain as well as tectonic terrain. At the top of the plateau are similar terrain types, seemingly continuous with those around the base. As with most of the uplifts we have mapped, these properties do not seem to fit the idea of convective upwelling in thick ice.

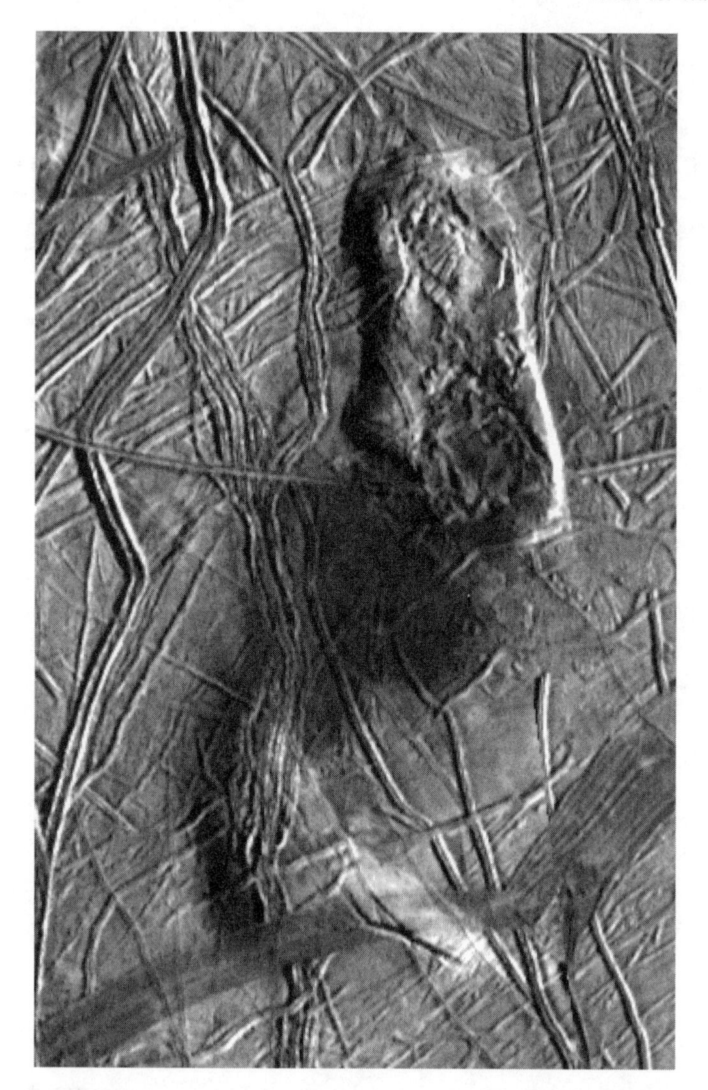

FIGURE 14.3: Two of the largest uplift features identified on Europa, located just south of the equator in the trailing hemisphere (c.f. **Fig. 10.2**). This image spans about 100 km from top to bottom. The largest (more northerly) uplift is about 40 km long (north to south). The more southerly one is a bit smaller, and has an extension to the lower right that is cracked, very unusual for a Europan uplift feature. Claims that such cracks are common proved to be incorrect.

The results of Martha's survey allowed us to look at the statistics of the sizes of pits and uplifts. We already knew that the claim that most PSDs are 10 km across was not true, because most are chaos, and chaos does not have such a characteristic size. But could it be that the pits and uplifts have a typical size that might support the thick-ice model? Martha's numbers showed that, like chaos, the pits and uplifts become more numerous the smaller their size, right down to the smallest features recognizable, given the limiting resolution of the images. There is no indication of a typical size for either pits or uplifts. The actual distributions of feature sizes, as well as the distribution of spacing, do not seem to be consistent with models of convection.

One interesting statistical result was that the distribution of sizes for pits and uplifts was very similar in the northern leading hemisphere and in the southern trailing hemisphere. There is a slightly different distribution in the southern leading hemisphere and the northern trailing hemisphere (relatively more in the 5-to-8-km size range). In other words, there is a sort of diagonal symmetry. It is as if the rocks in Africa and North America were similar to one another, while a different kind of rock was found in both South America and Asia. What could regions on obliquely opposite sides of the globe have in common?

The answer may be connected with polar wander. Remember, Alyssa's study of strike-slip faults showed that the ice on the leading hemisphere used to be further north and the ice on the trailing hemisphere used to be further south. So the pits in both the southern leading hemisphere and the northern trailing hemisphere used to be near the equator, while the other areas Martha studied used to be near the poles. Perhaps different amounts of heating at the equator explains the different size distributions in some way. In any case, the oblique symmetry seems to fit with polar wander. We found a similar, but less pronounced oblique symmetry in the statistics of uplifts and chaos, which may also support the evidence for polar wander.

Our study of pits and uplift features showed that the earlier taxonomy of "pits, spots, and domes" was premature. Properties of PSDs that had been widely cited as primary evidence for convective upwelling in thick ice were not supported by consideration of the available data, and they are most emphatically not supported to this very day. Yet those preliminary generalizations were repeated and propagated extensively, and widely accepted as fact.

With a more accurate description of the features on Europa, we can summarize what was wrong with the description in the PSD story. Most PSDs are indistinguishable from areas of chaotic terrain, a widespread phenomenon that covers a substantial portion of Europa. Patches of chaotic terrain include a continuum of sizes from more than 1000 km across, down to the smallest sizes recognizable in available images. Most of the PSDs are simply examples selected from a narrow range in that broad size distribution. The appearance of a characteristic size was an artifact of the selection of only features in that narrow size range as type examples of PSDs.

The confusion was exacerbated by the incorrect application of the word *lenticula* to include all of the PSDs. *Lenticulae* had been formally defined as spots of a certain size, and most of them are associated with very small patches of chaos. Lenticulae are not all related to pits or uplifts, which are not normally dark. And *by definition* they are not related to larger patches of chaos. Nevertheless, the repeated misuse of the word in the official mantra of Europa facts supported the impression that the vaguely defined classes of features called PSDs were all of the size defined for lenticula.

The origins of pits and uplifts on Europa remain uncertain, as do their implications for the bigger story of the structure and history of the satellite. The fiasco of "pits, spots, and domes" resulted from prematurely concluding what they meant before we knew what they were like. An accurate description and characterization of the population of these features should have been the first step toward their interpretation.

Once we surveyed these features and published our results, an interesting thing happened. The obligatory summaries of facts about Europa downgraded PSDs from established fact to a newly described category called "controversial pits, spots, and domes," definitely a small step in the right direction. But then, just as we began to hear less and less about "pits, spots, and domes", the defenders of thick ice introduced a completely different argument, this time based on the properties of impact craters, which I discuss in the next chapter.

15 The Scars of Impact

Impacts by projectiles that have crashed into Europa have not had very much effect on the surface we see today. However, they do provide some critical information about what has gone on there. The numbers of impact features indicate how long the surface has been in place. Specifically, the paucity of craters suggests a young surface. The scars of impacts also depend on what lies below. On Europa, the character of impact sites may reflect the distance down to liquid water.

In order to translate the number of Europan impact craters into information about the surface age, we would need to know the bombardment rate. If the rate had been low, even a very old surface might have few craters. The good news is that the numbers of craters are so very small that, even with only rough estimates of bombardment rates, we can be confident that the surface is extremely young compared with the age of the solar system. The other good news is that astronomers have made great progress in understanding the bombarding populations (especially of small asteroids and comets), which will improve the accuracy of surface-age estimates.

Like the measurements of age, measuring ice thickness using impact features requires information about a variety of phenomena. We need to understand what types of scars would be produced by impacts into ice of various thicknesses. Studies of the Earth, Moon, and other terrestrial planets have taught us a great deal about crater formation on solid bodies, but in order to interpret the record on Europa we need to know what happens when an ice crust, floating over liquid water, is bombarded. What sort of surface feature will result, and how does it depend on the thickness of the ice and the size and speed of the projectile? Based on careful experimentation and computer modeling, planetary scientists can now make meaningful predictions of what sort of impact feature to expect under various circumstances. Even so, comparison with observed craters will be somewhat subjective. Moreover, whatever is learned about the thickness of ice will only apply to a specific place and time of impact, and impact sites are few and far between on Europa.

***○ *

The yardstick for measuring the age of Europa's surface has been developed by a team assembled by Kevin Zahnle, an atmospheric scientist at NASA's Ames Research Center. A crucial part of the development involves understanding the rates at which the orbits of comets, which are believed to be the dominant impactors on the Jovian system, can evolve onto collision courses with Europa. Zahnle wisely recruited as partners Luke Dones and Hal Levison (of the Southwest Research Institute), members of an international community of celestial mechanicians who have been revolutionizing our understanding of the dynamics of small bodies in the solar system.

A few years ago, Brett Gladman, another celestial mechanician, gave a presentation to a broader audience of planetary scientists where he said that, when he does research, he gets exactly the correct answer. That comment did not sound modest, but as a celestial mechanician myself, I understood what he meant. If a problem is well defined, so that you know exactly where a population of bodies starts in the solar system and on what orbits, you can compute where the material will go with remarkable precision.

If we know exactly where comets are, and how many there are, and how big their solid icy bodies (their *nucleii*) are, and how their numbers vary with their sizes, and how big a crater forms from a given impact, we should be able to know how long it takes to make the numbers of craters on Europa. But, even though Zahnle's *ad hoc* team had the ability to do precise orbital computations, they first needed information about the population that supplies the impactors. They considered astronomical observations of the small icy bodies in the outer reaches of the solar system, they considered the statistics of comets that have moved in among the planets, and they accounted for astronomical and spacecraft studies of the sizes of nucleii of comets. They also needed to consider the size distributions of craters on the other moons of Jupiter to get an idea of how the numbers of impactors vary with size. And they needed to take into account our best understanding of crater-formation processes.

By carefully integrating current best understandings in all these areas, Zahnle estimated in 1998 that the numbers of craters on Europa implied the surface was only 10 million years old. They also appreciated that, no matter how exact the computations of celestial mechanics may be, there is considerable uncertainty in the other aspects of the problem, so they allowed for an uncertainty factor of 5 in the age. By 2003, new data had shown that the numbers of small comets is probably smaller than the earlier estimate, so the age estimate was revised upward to ~50 million years, consistent with the earlier estimate of the amount of uncertainty.

To put the age of 50 million years into perspective, we can consider events on Earth in the same time frame. Dinosaurs became extinct about 65 million years ago. Any astronomers among the last dinosaurs would have seen a completely different Europan surface from what we see now. It would have had similar *types* of features produced in the same ways: tidally driven tectonics would have formed similar patterns, but the specific features would all have been completely different. Moreover, the continual reprocessing of the surface means that the Europan surface at the time of the last dinosaurs would have been completely different from the time of the early dinosaurs. The Earth's surface has been changing quickly, as well; 50 million years ago the Atlantic Ocean was only beginning to open up, separating the Americas from Europe and Africa. The fact that both Earth's surface and Europa's surface change fast explains why neither planet features large numbers of craters: in the continual process of resurfacing, craters vanish as new surface replaces old.

The paucity of craters implies that the surface must be young, but it does not mean that a global resurfacing event wiped the slate clean. The surface appears to have been continually renewed by chaos disrupting previous terrain, by ridges overlying it, and by dilation and convergence. So the age given by the crater record represents the time-scale over which these gradual renewal processes have been continually replacing the surface.

The smallest Europan impact features are perfect little bowl-shaped craters. Many, smaller than a kilometer across, are visible in high-resolution images, and especially obvious where the surface is smooth enough to see them easily. For example, plenty of them are apparent in the smooth pull-apart bands of Astypalaea (**Fig. 15.1**). Most of the tiny craters are "secondary"—that is, formed from the spray of ejected ice from larger impacts. In Conamara, for example, they tend to be in the same portion covered by the bright ray from Pwyll crater (see **Color Plate 3**), and probably mark part of the same spray of material.

Larger craters are sufficiently few and far-between that none were captured in the Galileo mission's limited number of high-resolution images. Much more of the surface was covered by lower resolution images, but the relatively poor resolution limits the sizes of craters that can be recognized. Unfavorable lighting conditions make it even harder to find craters in many of these images. Nevertheless, a couple of dozen craters with diameters of 1 to 5 km have been identified, and they all have the simple bowl shapes expected for impacts by small projectiles into a solid body. These impacts probably probed to less than a couple of kilometers, so we cannot tell from the craters' bowl shapes whether the the ice is 2 km thick or thirty. But these

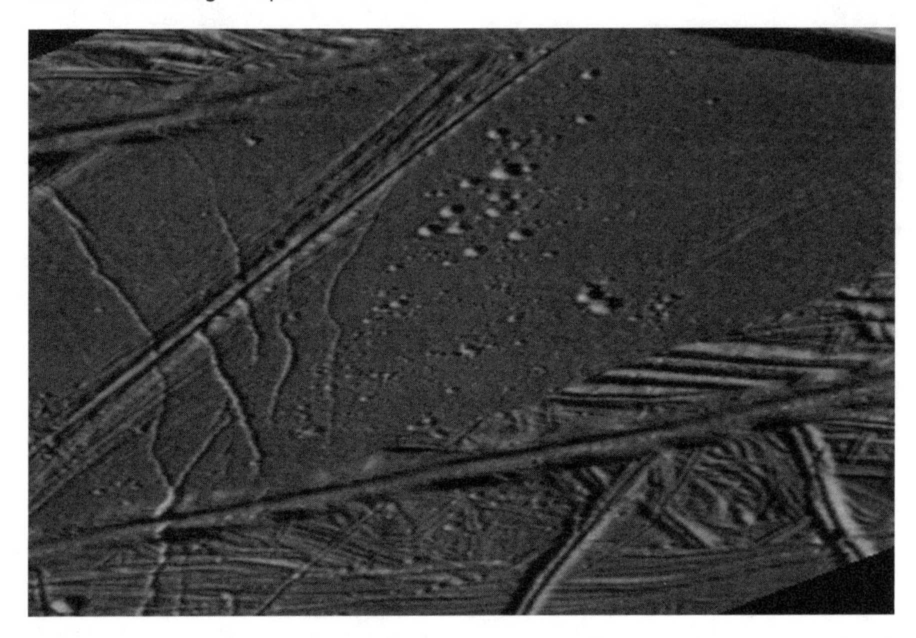

FIGURE 15.1: A cluster of small, bowl-shaped craters is readily visible on one of the smooth areas of Astypalea. This high-resolution (40 m/pixel) image is an oblique view, so the craters look elongated. The largest craters here are about 1 km in diameter. This cluster was probably formed by material sprayed from a larger impact nearby.

crater shapes do fit our estimates of tidal heating: The ice is at least a couple of kilometers thick.

The largest impact sites are Callanish and Tyre (**Fig. 15.2** and **Color Plate 6**), which happen to lie nearly opposite one another on the globe of Europa. These features are enormous multi-ringed structures. The inner rings are over 40 and 50 km across, respectively, and other rings extend far beyond them. The multi-ring form represents the damage that was done as the shock wave from the impact propagated outward from the center.

In describing Tyre, the imaging team compared it with multi-ring basins on other planets and with similar-sized impact scars on neighboring Ganymede. Such comparisons have little meaning, however, because any impact that created a circular structure so big would have penetrated at least 30 km through solid ice, according to well-understood principles of impact mechanics. Even the advocates for the canonically approved thick-ice model have generally been describing ice thinner than that. So detailed comparisons with the appearance of large impact structures on solid planets are probably irrelevant at best and misleading at worst. A more relevant comparison might have been found by looking for terrain similar to Tyre on Europa itself. And as it turned out, you did not have to look far.

FIGURE 15.2: The multi-ring impact feature Callanish. The impact that created Callanish punctured the ice, leaving a surface appearance similar to chaotic terrain. The area shown here is about 50 km by 125 km.

The appearance of the center of Callanish or Tyre is nearly indistinguishable from the chaotic terrain so common on Europa. At these impact sites we see the usual characteristics of chaos: the rafts of older crust; the lumpy, bumpy matrix; and even, on Tyre, a recent crack with incipient ridges crossing the refrozen matrix. Aside from the concentric placement of the larger rafts, this terrain differs from other chaotic terrain only in its setting within major impact features.

Tyre and Callanish record impact penetration to liquid water, which resulted in temporary exposure of the ocean to the surface. Blocks of crust were broken apart and dislodged, then locked into place as the exposed water refroze. From the perspective of the melt-through model of chaos formation, it is no surprise that this terrain looks so similar to more typical Europan chaotic terrain. The only difference is that in one case the crust was open to the ocean by an external impact, and in the other case the crust was opened by heat from below. For the impacts, the rafts form concentric patterns, but otherwise the basic character of the terrain that is produced in the two cases is indistinguishable, because the basic process is the same: temporary exposure of the liquid ocean at the surface, followed by refreezing.

In fact, Greg Hoppa came up with an amusing way to test the similarity of these impact sites to ordinary chaotic terrain. I thought it was hilarious. He showed pictures of the centers of Tyre and Callanish and other large craters to various people engaged in Europa image interpretation, and all but one of them identified these places as chaotic terrain. Jeannie Riley, our undergraduate assistant who had done the definitive chaos survey, was the only person who could distinguish between images of chaos and of impact penetration.

There were reasons it was hard to make the distinction. Even the darkening and coloration are similar. At low resolution—all that was available in Voyager and early Galileo images—Tyre was indistinguishable from the splotches that represent chaotic terrain. For example, in global images like the ones we saw in **Chapter 2**, the large impact site Tyre and the large chaos areas Thrace and Thera all appeared as similar dark areas—similar enough in fact that all were given the same designation "macula" (dark area) by the IAU nomenclature gurus. That time, by chance, the official name-givers got it right: Thrace Macula, Thera Macula, and Tyre Macula have more in common than not, even though one is a major impact feature. The darkening that gave them all the appearance of maculae resulted from a fundamental similarity: They all represent penetration to the liquid ocean.

The similarity of large impact features to chaotic terrain is a very big problem for the canonical thick-ice, isolated-ocean model. And applying the same designation, *macula*, to both the impact and chaos features only emphasized the similarity. So I was not surprised when the nomenklatura of nomenclature, under the control of the thick-ice advocates, quickly renamed Tyre Macula to just plain Tyre. Keeping impact features and chaos features distinct, at least in name, helps defend the thick-ice canon.

***○ *

We have seen that the strongest impacts into Europa's surface poked right through the ice, leaving complex features like Tyre and Callanish. The ice must be thinner than ~30 km to have allowed such penetration. On the other hand, more minor impacts did not penetrate through the ice at all. Small craters, up to about 5 km in diameter, have simple bowl shapes, just as expected if the target is solid material. Those small craters tell us that the ice there must have extended at least a couple of kilometers below the surface. So the craters we have considered so far—the really big complex ones and the small simple bowls—bracket the ice thickness somewhere between ~2 and ~30 km.

That range is too broad to address the key issue of the permeability of Europa's ice crust. In order to constrain more precisely the thickness of the ice, we need to consider craters of intermediate size (bigger than the simple bowls and smaller than Tyre and Callanish). The bad news is that there are few in that size range, so a systematic size-dependent trend in their character may be hard to find. A larger sample would have been better because any quirky properties of individual impact sites would average out, making meaningful trends more apparent. The paucity of mid-size craters is not all bad news though: With so few craters it is easy to get to know them all.

I will describe a few of these critical mid-size craters, but here is the main thing you need to know: The floors of most craters larger than about 10 km in diameter consist of terrain that is indistinguishable from chaos.

Consider the 19-km-wide crater Amergin, at the upper left edge of **Fig. 15.3**. Amergin happens to be near the edge of available imaging. With slightly different camera pointing, we would have had one fewer mid-size crater in our sample. Also, the shadows are unusually long, because this picture was taken near sunrise. Unlike Tyre and Callanish, Amergin has a distinct, single, upraised rim. But inside the rim, the terrain is indistinguishable from chaotic terrain elsewhere, as for example in the large patch at the lower center portion of **Fig. 15.3**. **Fig. 15.4** demonstrates this similarity further. Like Tyre, Callanish, and chaos everywhere, Amergin shows up as a typical dark spot when imaged at low resolution. But on closer inspection, the surface reveals itself not as a smooth, dark grey blur, but as a lumpy jumbled refrozen mass. Despite the fact that Amergin is far smaller

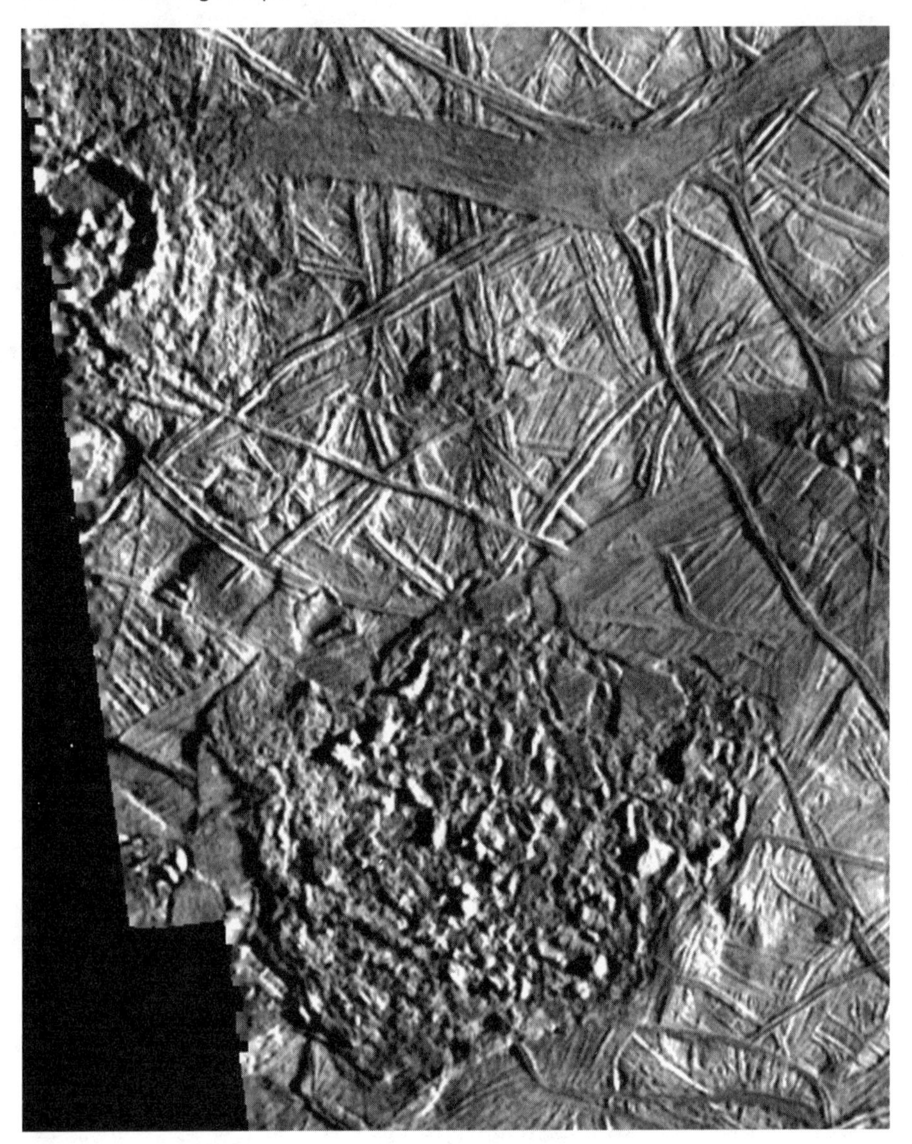

FIGURE 15.3: Amergin crater, at the upper left edge of this image, has an interior indistinguishable from chaotic terrain. For comparison see the nearby but much larger patch of chaos at bottom center. Amergin's floor, from rim to rim, is about 19 km. This image comes from the 200 m/pixel Regional Mapping sequence in orbit E17.

than Tyre or Callanish, its appearance is consistent with penetration to liquid.

Of similar size to Amergin, with a diameter of about 21 km, is Crater Manannán (**Fig. 15.5**). Manannán has a less pronounced rim, but again the

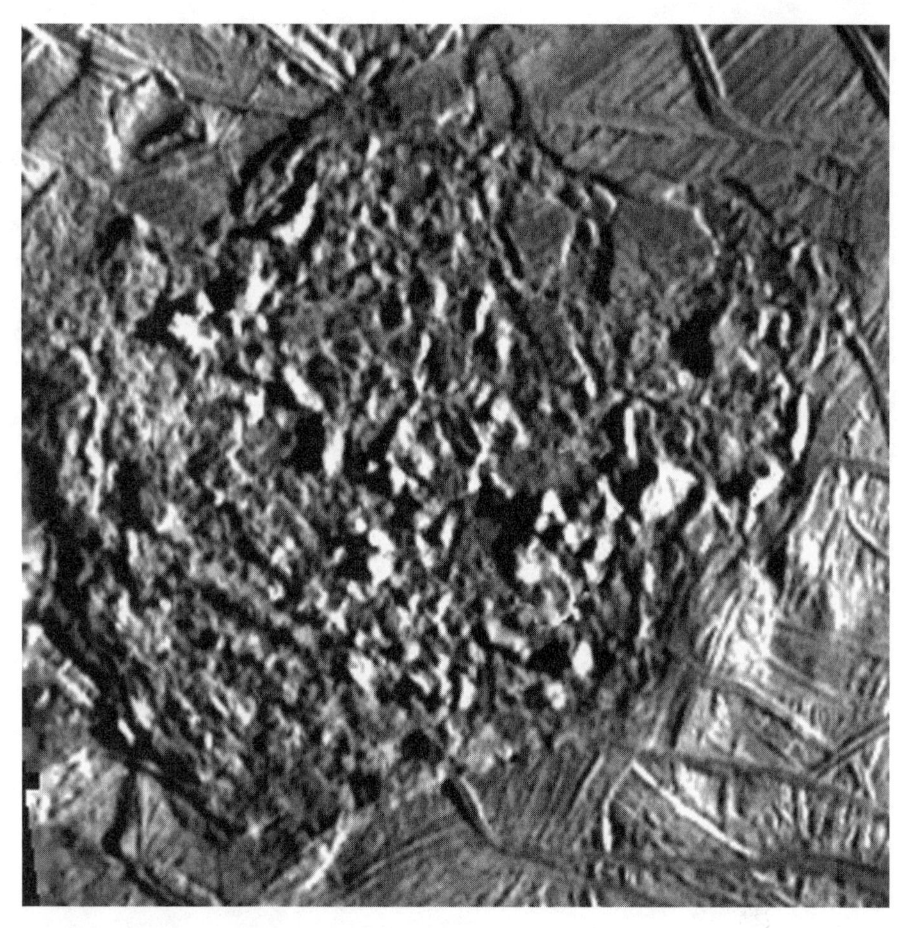

FIGURE 15.4: Where's Waldo? Amergin's interior hides in plain sight in the similar terrain of nearby chaos. In this manipulated image, the floor of Amergin has been pasted into the larger nearby chaos (the true appearance of which can be seen in the lower central part of Fig. 15.3).

interior floor looks like chaos. Also, like chaos and other large craters, Manannán appeared as a dark spot in global-scale images. Extending roughly 10 km beyond the rim of the crater is a lumpy blanket of ejected material. At low resolution a relatively bright halo extends 70 km outward, probably caused by a spray of fine ejecta, similar to the bright rays from crater Pwyll (**Color Plate 3**).

The main difference between Amergin and Manannán is that one impact occurred in tectonic terrain (**Fig. 15.3**) and the other in a pre-existing chaos area (**Fig. 15.5**). But within each crater, we see chaotic terrain, consistent with penetration through to the ocean.

Crater Pwyll is one of the most obvious impact features because of its

FIGURE 15.5: Crater Manannán is about 21 km in diameter. It lies in a region rich in chaotic terrain (including the nearly 20-km-wide patch at the left in this image). The interior of Manannán is, like all craters larger than 10 km, similar to chaotic terrain. This image was taken during orbit E11 at 218 m/pixel.

prominent bright rays. That the rays still stand out so prominently suggests that it may be one of the most recent, as well. Otherwise, 24-km-wide Pwyll is fairly typical of the other mid-size craters we've examined so far (**Fig. 15.6**). Like the others, its interior lumps and bumps and rafts are similar to sites where the ocean has likely been exposed. And, as at other sites of ocean exposure (like chaos), the interior of the crater is dark with the dark coloring extending several kilometers outward around it (as shown in **Color Plate 7**).

With its bright, well-preserved rays extending dramatically from its center, Pwyll Crater presented such an arresting appearance that it was targeted for better imaging relatively early in the Galileo mission. Not surprisingly, interpretation of its appearance became the topic of considerable debate within the imaging team, especially because of its implications for ice thickness. I myself followed the discussion with interest, but detachment. For one thing, no one was interested in my opinion. Also I was only beginning my work on tidal tectonics. I never imagined that my

FIGURE 15.6: Pwyll Crater is about 24 km across. As expected, the interior has the appearance of chaotic terrain. This image was taken during orbit E6 at 240 m/pixel.

work would lead me to become the "standard bearer" (or maybe "fall-guy" is more appropriate) for thin ice.

Establishment of the party line on Pwyll was the same as the process followed for Conamara Chaos, as described in **Chapter 13**. Remember, Conamara had been initially viewed as a site of oceanic exposure, but then the thick-ice proponents exerted considerable pressure to ensure that the authorized position of the team favored a formation process based on thick, solid ice. That rewriting of Conamara's history took place over the course of 1997, from the first revealing images (**Fig. 2.7**) taken during orbit E6 in February to the time that early reports were due to be produced later in the year. In exactly the same time frame, Pwyll got the same treatment. Pwyll's similarity to chaotic terrain was recognized in its orbit E6 image (**Fig. 15.6**), and the lumps, bumps, and rafts in the crater were hypothesized to be pieces of crust that had floated into position. In fact, some of the rafts within the floor of the crater were thought to have come from the notches visible in the rim (**Fig. 15.9**). But then, just as for Conamara, the politically powerful and aggressive thick-ice advocates quashed that hypothesis. By the end of the year, the imaging team's papers on craters made no mention of any similarity between chaotic terrain and the interiors of impact features.

The canonical approach does not consider similarities between craters and other terrains on Europa. Instead, these craters, with characteristics unique to Europa, are forced into categories that had been developed to describe craters on solid bodies. And once you describe Europa's impact features in terms of classes of craters seen on solid planets, theorists who model their formation are likely to conclude that the impacts must have penetrated only into solid material. In other words, if you assume that the ice is thick, then the craters on Europa prove that the ice is thick. The logic is blatantly circular, but after the PSD mythology had lost its credibility, craters were adopted as the evidence *du jour* for the canonical thick-ice theory. Interestingly, as with the PSDs, this argument for thick ice depended on a false taxonomy.

Despite Pwyll's similarity to chaotic terrain, it was categorized as a "central peak crater," which is a well-defined standard class based on a long tradition of crater studies. On solid bodies, small craters are usually simple bowls, while the products of more energetic impacts are "complex" craters, with features like central peaks and flat floors. Central peaks are created immediately after impact excavation, as some material flows back toward the center, and rebounds upward. Central peak craters have been examined in great detail in their natural settings on solid planets, and they have been modeled using impact experiments, numerical simulations, and analytical theory. Their formation is very well understood.

So, just what is Pwyll? Given only a choice between (a) a "simple crater" or (b) a "central peak crater," then (b) is undeniably a closer fit. But that extremely limited pair of options does not take into account the resemblance of the interior terrain to chaos, and the similarity of the "peaks" in Pwyll to the rafts and ice floes of chaos. It disregards the random placement of these "peaks," which could hardly be described as "central." Limiting the choices to (a) and (b) ignores the fact that Europa is completely different from any planetary body we have observed before. Categories of craters identified elsewhere may not cover all the possibilities, such as penetration to an ocean, that are possible on Europa. Forcing Pwyll or any of the other mid-size craters into one of these limited categories precludes consideration of formation mechanisms that may be unique to Europa.

A few of these craters do have their largest lumps concentrated toward the center, but that geometry is perfectly consistent with their similarity to chaotic terrain. It is reasonable that in some cases breakaway crustal rafts have sloshed toward the center during in-fill of the initial crater.

The authorized descriptions of craters by the Galileo imaging team use phrases like "central peak complex." In a cultural vacuum, I would be content to accept that terminology, but in the context of our ongoing disagreement, this language conveys the notion that the crater formed like other central-peak craters on other planets—by impact into solid material. Instead, the bumps could be described as "crustal rafts displaced toward the center of the crater." That terminology would be more consistent with the appearance of most craters on Europa, and admits the possibility that the character of these craters results from processes very different from what would have happened on a solid planet.

Even among advocates for the traditional central-peak interpretation, there is disagreement about which craters fall into that class. In an influential article in *Nature*, Paul Schenk described all 20 known craters in the diameter range 3–30 km as being central-peak craters, indicative to him of formation in completely solid material. Bear in mind that this group of about 20 craters includes Amergin, Manannán and Pwyll (**Figs. 15.3, 15.5, and 15.6**). Evidently central peaks are in the eye of the beholder, as is blindness to the similarity to chaos.

Although Schenk saw all twenty of those impact features as central-peak craters, Zibi Turtle and Betty Pierazzo wrote in *Science* that only six of them are. Like Schenk, they included Amergin and Pwyll, but at least they had the good sense to dump Manannán. Of their six "central peak craters," only four have actual peaks near the center.

Only a few craters have central bumps, and all craters bigger than 10 km have floors that look like chaotic terrain. Yet the canonical position is that all of these craters formed in solid material, and none penetrated through the crust.

In order to form central-peak craters by the standard mechanism for solid bodies, the ice on Europa would need to be thick enough to prevent interaction with the ocean. In their *Science* article, Turtle and Pierazzo described computer simulations of the formation of 20-km-wide craters. Eventually they found that the ice would have to be thicker than 12 km for central peaks to form in the usual way, in solid material without involvement of the underlying ocean.

In the same issue of *Science*, the magazine's staff writer Dick Kerr reported that Turtle and Pierazzo's work put "a lid on life on Europa." He disparaged my results regarding ocean-to-surface connections, invoking the standard mantra of the "pits, spots, and domes" to support the idea of thick ice. Had Kerr tried to obtain my view of the story before editorializing for the party line, I would have interpreted the crater modeling differently for him.

The kind of quantitative modeling that Turtle and Pierazzo had done is an essential part of the process of understanding Europa. However, a chain of logic is only as strong as its weakest link. The numerical modeling of impacts is a strong link. But the chain of logic that connects a few strange craters to the thickness of ice depends on a belief that typical 20-km craters (like Amergin, for example) look so much like central peak craters on solid planets that they *must* have been formed the same way. That belief is a very weak link.

***O *

We have already seen how the justifications for the thick-ice, isolated-ocean model of Europa appeared not in regular scientific journals, where all the details of a case are made, but in abbreviated form in the magazines *Science* and *Nature*. Remember, the entire case for solid-state convection was made in a couple of paragraphs in a letter to *Nature* about the so-called "pits, spots, and domes." Similarly, the case for surface corrugations requiring thick ice was made in just a couple of paragraphs in another letter to *Nature*. Then, the crater modeling was published as two pages in *Science*. No reasonably inclusive amount of detail can be included in such short papers, yet these results are accepted as the fundamental supporting evidence for the thick-ice, isolated-ocean model because *Science* and *Nature* are viewed as the voice of scientific authority.

The isolated-ocean advocates finally toned down the "pits, spots, and domes" mantra after we showed that almost nothing that had been said about them was true. At that point, the thick-ice advocates turned to the craters as their principle argument. Dick Kerr of *Science* really helped them by setting up that new spin on the party line.

The thick-ice advocates must have also welcomed a more recent interpretation of the impact record by Paul Schenk. Again, the documentation is sketchy, only two pages in a letter to *Nature*, but it gives the appearance of densely packed technical detail, with a definitive conclusion that the ice must be thicker than 19 km. The use of two-digit precision packs a lot of authority. Because few people bother to examine the details, such papers make powerful and persuasive cases for thick ice.

A closer look leads to very different conclusions. Schenk's story depends on the depths and diameters of craters. His measurements of depths are questionable, but for this discussion I will assume that they are acceptable, because there are enough other problems with his story. Schenk plotted the depth and diameter of each crater on a graph (**Fig. 15.7**). Each circle represents one crater on Europa. Black circles are simple craters, open circles are "central peak craters," the circles with a + inside are the "modified central peak craters Mannann'an [*sic*] and Pwyll," and the long "I" symbols are the multi-ring features Tyre and Callanish. The classification of so many as "central peak craters" is unique to Schenk.

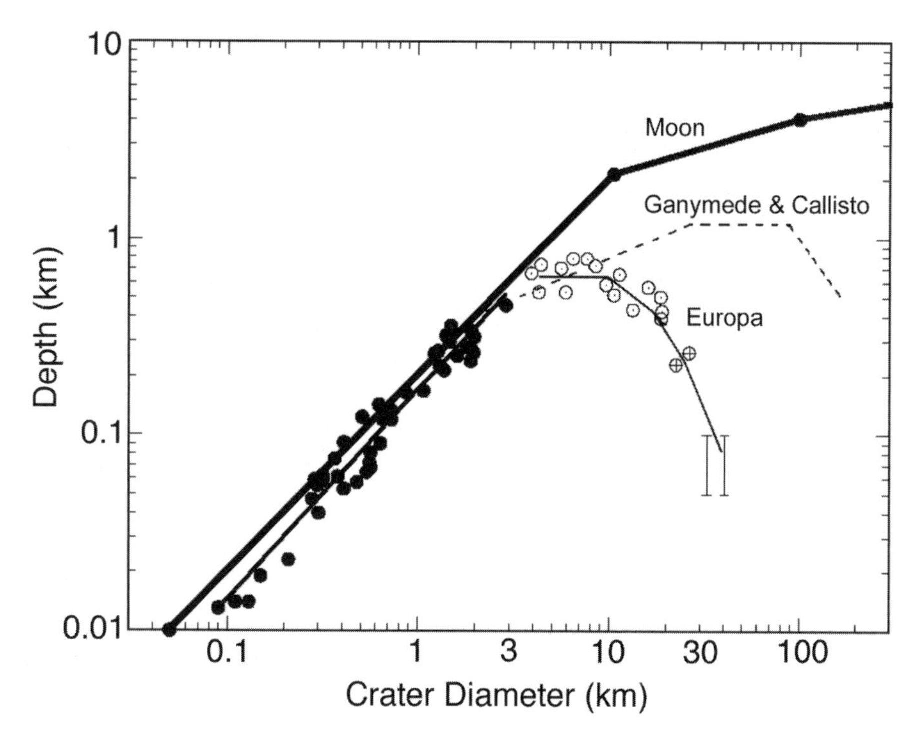

FIGURE 15.7: Depths, diameters and classification of impact features are given by the symbols plotted by Paul Schenk: Black spots represent simple craters, circles with central dots represent central peak craters, circles with crosses represent "modified central peak craters Mannann'an (*sic*) and Pwyll," and the two vertical bars represent the diameters and range of possible range of depth of Tyre and Callanish. Classifications are subjective: Only Schenk identified so many as "central peak craters". The heavy solid line is Schenk's description of the depth/diameter relationship for craters in a solid body, the Moon. The dashed line shows the trends of Schenk's results for Ganymede or the nearly identical ones for Callisto, and a fine solid line follows the trend of his data points for Europa. These satellites are very different from the Moon. The transition to impacts that penetrated to liquid water on Ganymede and Callisto, according to Schenk, is shown by the bend at crater diameters of 100 km. Strangely, Schenk believes that the Europa curve at the 30-km diameter is similar to the Ganymede and Callisto curve at 100 km. That is the basis for his case that the ice on Europa is thick.

The thin solid line is how Schenk connected the dots. For comparison, the dashed line shows the depth-to-diameter relationship for Ganymede and Callisto, the two other icy Galilean moons of Jupiter. The heavy solid line is for craters on Earth's Moon.

For Ganymede and Callisto, the dashed line bends at 100 km. The depth suddenly drops for impact features larger than 100 km in diameter. To explain this observation, Schenk suggests that the larger impacts penetrated to a layer of liquid water about 80 km down. Such speculation has some

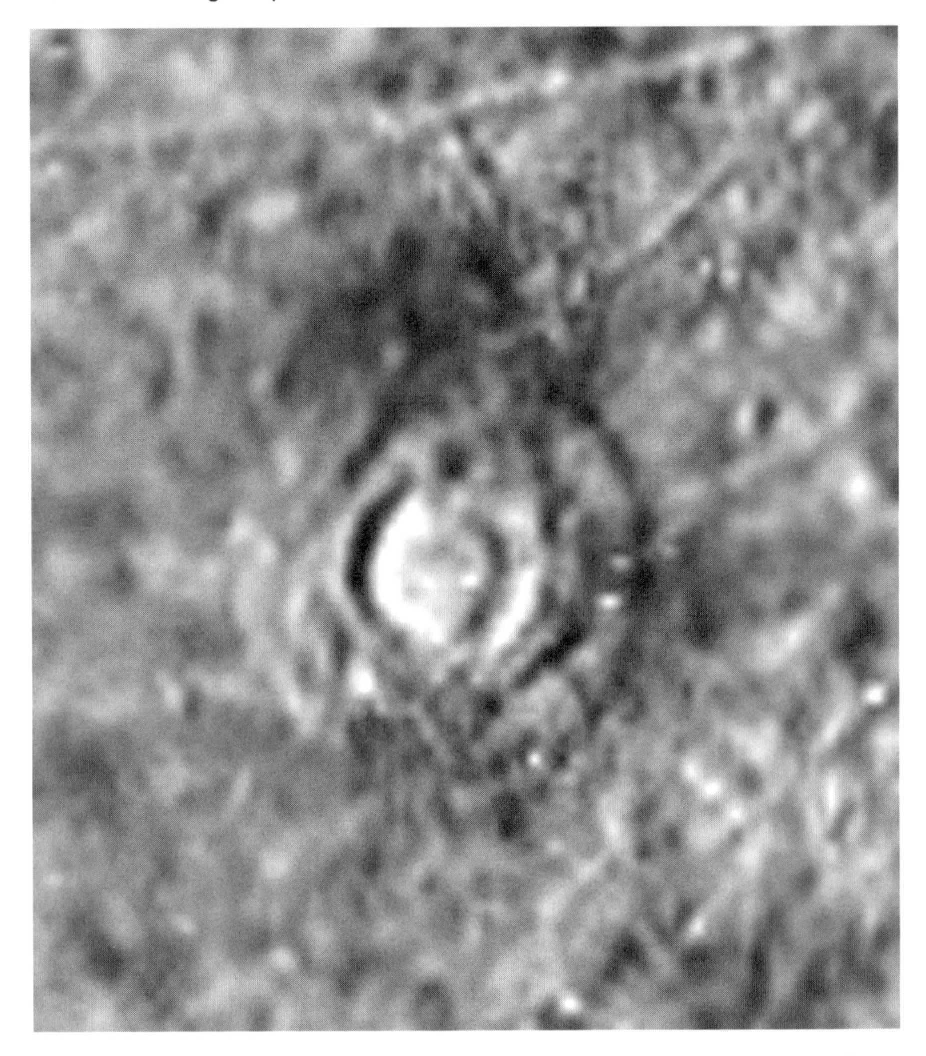

FIGURE 15.8: The best picture of crater Tegid is from a global-scale sequence taken during orbit E14 at 1.4 km/pixel. The central floor is about 20 km across. Despite the low resolution, a multi-ring structure is visible around this relatively small crater. The similar-sized Taliesin is also a multi-ring crater.

merit, because the pressure at such great depths might be adequate, combined with the modest heat sources in those satellites, to produce at least a thin liquid layer within those moons.

For Europa, Schenk believes that the analogous transition occurs at 30 km diameter. Apparently, to his eye, in **Fig. 15.7** the same "sharp reduction in crater depths and development of anomalous impact morphologies" represented by the Ganymede and Callisto line for diameters of 100 km and greater is seen on the Europa line for diameters of 30 km and greater. So he concludes that only craters larger than 30 km represent impacts that penetrated through the ice to the liquid, implying that the ice must be at least 19 km thick. To buy that story, you would have to agree that in **Fig. 15.7** the shape of the Europa line at diameter 30 km is similar to to the Ganymede-Callisto line at 100 km. I found that claim bizarre, but Schenk became the darling of the thick-ice zealots, especially when their earlier lines of evidence collapsed.

When I look at the Europa line in **Fig. 15.7**, the downturn occurs at diameter 10 km, not 30 km. Using Schenk's reasoning, then, craters as small as 10 km penetrated through the ice. And, consistent with that result, all craters larger than 10 km look like chaos. Schenk's own evidence indicates ice thinner than 10 km.

Moreover, Schenk may not have known about two craters, Taliesin and Tegid (**Fig. 15.8**), which were only imaged at very low resolution. They are comparable in size to Pwyll and Manannán, but seem to have multi-ring structures. Again, forms that Schenk attributes to penetration to liquid appear at smaller sizes than he admitted.

Schenk's *reading* of his graph perfectly fit the party line that the ice is thicker than 20 km. The conclusion of his letter to *Nature*, which is all that most readers scan, tied this result from the impact record to the conventional evidence for solid-state convection, the well known "ovoid features". Egg shaped? I assume he meant the PSDs, which were supposedly (but not really) roundish. Never mind that a reasonable reading of the impact record, including Schenk's own data, suggests that the ice is thin, not thick; never mind that the observational evidence cited for solid-state upwelling had no basis in fact no matter how frequently it was repeated; never mind that his own graph undermines the conclusions he draws from it. If you want to get along, believe in thick ice.

After Schenk's *Nature* paper, there has been less talk of craters as evidence for thick ice. I like to think that people thought more about it and realized the weakness of those arguments. Whatever the reason, the campaign for thick ice shifted to other, equally shaky evidence. Among the thick-ice advocates, the conclusion seems fixed while the arguments for it keep changing.

The Bandwagon

They were no less ridiculous than those who in great number opposed my first celestial discoveries, persuading themselves (as is usual in noisy altercations of idle words) that by texts, authorities, syllogisms, and their foolishnesses they could force the course of nature to conform to their dreams. Malignity, envy, and ignorance are unconquerable beasts, and I see by daily experience that my contradictors, though overthrown by a hundred confrontations and past experiments, and made certain that new opinions introduced by me and at first denied by them are true, do not cease to oppose themselves to other things day by day propounded by me, still hoping some day to have me on the hip and that some one small error of mine will cancel all the other true teachings I have introduced. Now, you must let the vulgar shout, and just continue in conversation with the Muses, enemies of the tumultuous rabble.

—*Galileo, 1630*

IF someone of Galileo's towering intellectual stature had to put up with all that, it should come as no surprise that similar rough treatment has been common for scientific innovators ever since. I find the process disturbing, but in a way it is comforting to look back at scientific history and realize that bucking the canon does not necessarily mean being wrong. I cannot tell you how delighted I am that Craig Venter got so rich from human genomics after fellow scientists blocked his federal research funding. On the other hand, poor Alfred Wegener's success in revolutionizing geoscience came posthumously after a lifetime of derision for his ideas on continental motion. I am no Galileo (nor a Venter or Wegener), but the "tumultuous rabble" that he described has not changed a bit.

Something is wrong with the scientific enterprise as a whole if it routinely punishes innovation in this way. Astute scientists quickly realize that it is more rewarding in some very practical ways to stay within the mainstream. The peer-review system controls the fate of proposals for research funding and of manuscripts submitted to scientific journals. Being an insider also gets you invitations to membership on advisory panels that guide NASA's exploration planning. And letters of recommendation from recognized authority figures determine who wins the cutthroat competition for jobs.

Staying in the mainstream also ensures comfortable and friendly interactions at scientific conferences, whether during casual conversation or formal presentations. On the other hand, following scientific evidence where it leads, without regard to peer pressure, requires a thick skin or a willingness to be ground down emotionally—and tenure.

Being off the bandwagon was unhealthy for a career, especially for younger scientists. Greg Hoppa had made many of the key discoveries about Europa, knew more perhaps than anyone about the contents of the images, and had one of the finest minds among the new generation of planetary scientists, but he had no hope of a professional future that would support his growing family, so he left the field. Other young people, who toed the party line, were moving up in their careers, getting faculty jobs, serving on high-profile policy committees, and being honored for their contributions. The bandwagon for the convective-thick-ice isolated-ocean model had tremendous political momentum.

Young researchers interested in Europa faced a moral challenge. Doing good science was not necessarily the road to success, especially if their work yielded results that might be consistent with thin, permeable ice. On the other hand it was evident that, if they could skew their research to support the party line, they could expect to be rewarded.

And this situation persisted in spite of the fact that the thick-ice dogma was not faring very well against the facts. The standard mantra of pits, spots, and domes had finally come under scrutiny and, as the basis for that argument crumbled (Chapter 14), the crater story replaced it. But that line of argument proved to be equally flawed (Chapter 15). With the foundation of the party line needing shoring up, several papers appeared in the literature that tried to support the thick-ice model on different grounds. On closer consideration, these too used distorted logic to reach conclusions fitting the party line: the canonical thick-ice conclusion remains constant, even as the rationale keeps changing.

The most prominent current rationale for thick ice is based on the topography of Europa's surface. I guess I have myself to blame. When I first wrote in 1999 how buoyancy could explain the bulgy look of some chaos, I described a test of the theory. A politician would not have suggested a way to shoot down his or her own idea. In science, though, you are supposed to look for ways to test your own ideas. The idea is to find the truth. Remember, buoyancy could only bring the chaos up to the general level of the surrounding terrain. So here is the test: If chaos is bulged up higher, then something must have pushed it, such as the solid-state upwellings imagined by the thick-ice party.

Starting around 2002, Paul Schenk and Bob Pappalardo reported in presentations at various conferences that they had in fact measured the heights of chaotic terrain in various places. Indeed, they reported that it was distinctly higher than the surroundings, proving that the ice must be thick. Based on their past performance, I was not surprised that quantitative details of their method were missing. I remember asking after one of Schenk's talks whether his measurements of elevations were precise enough to support his claims. Schenk told the audience they were. Plenty of people believed him, and they let me know that the idea of thin ice was dead—again.

Two methods have been used for measuring the elevation of the surface. One uses stereo imaging. You can demonstrate this method for yourself. Blink one eye closed at a time, and go back and forth between the two. Things near you, like this book, jump left and right, while the background stays still. Even when both eyes are open, your brain compares the two images and (without you even knowing it) does calculations about how far things are. That is how most people see the world in 3D.

In principle you can do the same thing if you have two images of Europa's surface taken from two different angles, like the views from two eyes. The high spots are closer and the low spots are farther. If you do the math you can figure the elevation of things on the surface. This method is used a lot on Earth, where we can get plenty of really great, well-controlled stereo imaging from aircraft and satellites.

The other method that the thick-ice adherents reported using is photoclinometry: *photo* as in light; *clin* as in incline; and *metry* as in measure. The idea is that a uniform surface material will appear darker or lighter depending on the lighting angle and the tilt relative to the viewer. Without this effect a skier could not see moguls (large bumps in the snow); every day would be a white-out.

Schenk and Pappalardo in 2004 published maps showing elevations of selected areas where stereo images are available. Sure enough, the color-coded maps showed chaotic terrain bulging upward by hundreds of meters in flaming reds and purples above the yellows, greens, and blues of the surrounding terrain. The conclusion was that the melt-through model had failed the test, and that the ice is at least 25 km thick.

That report appeared in the journal *Geophysical Research Letters*, which, like *Nature* and *Science*, publishes only very short pieces. With all the colorful maps, there was too little space to detail the methods and assumptions involved, nothing beyond the statement that photoclinimetry had been used, along with some supporting stereo. We just had to trust that they made the measurements correctly to achieve the stated precision of "better than 10 m"—a very impressive achievement indeed.

The thick-ice bandwagon rolled on with this new fuel. I remained skeptical. Photoclinometry depends on knowing the exact optical properties of the surface. Slight variations in roughness, grain size, impurities, and shadowing can really mess up the results. And stereo requires that you recognize small features and measure exactly how much they jump sideways from one image to another. With pixel sizes of 200 m, how precise can you be? Show me the arithmetic!

I planned to look into the true precision of these techniques. But then something happened that saved me a lot of trouble. Two other thick-ice adherents, Francis Nimmo and Bernt Giese, published a similar paper, showing similar color-coded elevation maps. They discussed at length how the ice must be thick to explain the elevated chaos areas. They too claimed that their maps were precise to within 10 m in elevation, again with no supporting details.

What startled me was a strange remark in their paper, comparing their map of a region near the Tyre impact site with a map of the same area by Schenk and Pappalardo: "The two models are in remarkably good agreement." But if each is good to the nearest 10 m in elevation, why would it be remarkable that they agree? Even stranger, in comparing their maps, it was plain to see that the two sets of results did not agree at all. The flaming red (high) area of chaos in Schenk and Pappalardo's map is a blue (low) area in Nimmo and Giese's. Remarkably good agreement? Hardly. The results disagree by way over 100 m in elevation—a big error.

Actually, it is fairly easy to look at image pairs in stereo. You could use an antique Victorian stereoscope. Or display the two images together on a computer screen, one in blue and one in red, then look at them with 3D glasses. So I checked out that chaos. It is not high; it is low. It looks like a place where the ocean melted through.

Schenk and Pappalardo's description of this chaos is wrong in another way, too. They published a profile of the elevations that runs across the chaos and over the neighboring tectonic terrain. On the profile, they labeled the high part as the chaos. But what they label *chaos* actually includes the adjacent tectonic terrain, which is equally high. And, if the elevations in the chaos are corrected, the only part of the profile that is really at a high elevation is the tectonic terrain, not the chaos. Even more bizarre, they label a little dip in the elevation profile as a "moat" around the supposedly elevated chaos. But really, that dip is a small crater in the middle of tectonic terrain. It is literally hard to imagine that Schenk and Pappalardo actually looked at the picture.

Maybe there is something legitimate in the topography claimed by these authors. But the case depends on a detailed quantitative analysis. They have a lot of explaining to do, but explanations seem in short supply.

***○ *

It must be fairly easy to publish papers that support the party line, and it must be rewarding, as well. Most remarkably, in some cases, analyses that provided good corroborating support for thin, permeable ice were hidden between introductory sections that denigrated evidence for thin ice and concluding sections that paid homage to the thick-ice, isolated-ocean model. I want to describe a few examples because they are part of the current case for thick ice. More important, they demonstrate how a seemingly detailed, quantitative analysis can be twisted to conform to a predetermined conclusion, keeping the authors safely on the bandwagon.

The first example is another argument based on topography. My student Randy Tufts had pioneered a technique for probing the structure of the crust on Europa by studying how it bends down due to loads on the surface. He investigated, as part of his PhD dissertation, ridges whose weight appeared to have warped down the surface on either side of it. The results were consistent with thin ice.

Recently Francis Nimmo got together with Bob Pappalardo, driver of the thick-ice bandwagon, and Bernt Giese to consider what they interpreted as down-warping of the lithosphere in another location. I have doubts whether the feature interpreted by Nimmo *et al.* is real and, if it is, whether it represents a down-warping under a load. But let's give them the benefit of the doubt. Based on their reported computations, the ice must be in the range between 6 km and 35 km thick. In other words, their result was equally consistent with either the canonical thick-ice model or with our model, in which the ice is thin enough for the ocean to be linked to the surface.

But then things get Orwellian. As Nimmo and Pappalardo must know, few scientists will study the whole paper. Most scientists will read only the summary abstract of the paper, and perhaps scan the conclusion. The bottom line is "a uniform present-day shell thickness for Europa close to 25 km." Never mind that their actual results showed the ice could be as thin as 6 km; the authors needed to stay on the thick-ice bandwagon.

***○ *

Another paper supporting the prevailing dogma involved strike-slip displacement, the shearing motion I discussed in Chapter 12. Strike-slip has been a thorn in the side of the thick-ice people because our model of tidal walking, which fits the observed distribution of strike-slip displacement so nicely, requires thin ice. So Francis Nimmo and Eric Gaidos scored very big points when they reported on computer simulations of strike-slip in thick ice. They also came up with a bonus: In their model, there would be enough friction for substantial heating under the length of the crack, providing a rationale for Head and Pappalardo's story of warm ice rising to build up into ridges.

In their simulation, the cold brittle ice in the upper couple of kilometers is cracked, so plates can scrape past one another. In our studies of strike-slip, these plates were free to slide over the liquid below. In the thick-ice model, the plates have to move over warm viscous ice. There has to be a lot of resistance. You have to read the paper very carefully to find that Nimmo and Gaidos modeled the viscous behavior unrealistically, but there is a much more fundamental flaw in the simulations.

In their model, the strike-slip displacement really moves along. The plates shear past one another at about 10 cm/day—downright speedy for a plate of ice. The underlying ice is distorted so fast that a great deal of heat is generated, raising the temperature 66°C. (These results are impressively precise.) Conveniently, that heat is exactly enough, the authors find, to buoy up the ice under the crack and thus create double ridges.

As in terrestrial global economics, sometimes it is helpful to follow the energy. If there is so much heating going on, something must be pushing very hard. But Nimmo and Gaidos simply imposed on their computer simulation a predetermined speed of 10 cm/day. That is mighty swift, and it is difficult to imagine what could drive it against the resistance of all that viscous ice.

It turns out they got that speed from our calculations of tidal strain. But in our considerations of tidally driven strike-slip, the ice can move that fast only if there is no resistance because the ice is floating across the liquid below. In other words, our calculation of the speed depends on the cracks reaching the ocean. Nimmo and Gaidos imposed this rapid movement on a model where there should be tremendous resistance from 18 km of solid ice below the crack. If their thick-ice model were really driven by tides, the actual speed would be much slower, and the resultant heating would be negligible.

In their computer simulation of thick ice, then, they forced the plates to move as fast as we had found for thin ice. On that basis, they concluded that rapid strike-slip displacement is possible with thick ice. Nimmo and Gaidos arrived at their conclusions by classic circular reasoning. But their paper earned them places of honor on the bandwagon.

I first heard of a young scientist named Jason Goodman after he did theoretical work that provided strong support for our melt-through model of chaos formation, then presented it as evidence that our model was wrong.

Evidently, as a graduate student in oceanography at M.I.T., Goodman had connected with Geoff Collins, one of Jim Head's graduate students at Brown. Collins had written a paper disparaging melt-through. One reason he had for dismissing our view was that the amount of heat required to melt a hole in the ice seemed like a big number to him. But as discussed in **Chapter 13**, the amount of heat needed is just a modest increase over the global average. Collins had also argued that, even if heat come from local hot spots in the ocean floor, it would spread out before it got up through the ocean to the ice crust. But then, as I discussed in **Chapter 13**, Thomson and Delaney showed that the heat could rise in a narrow plume.

Later, as a post-doctoral associate at the University of Chicago, Goodman seemed to some colleagues in oceanography to be determined to disprove our model for chaos formation. He made assumptions different from those Thomson and Delaney made about the unknown conditions in the ocean, and he performed laboratory experiments to help understand the rising of warm buoyant plumes. What he found is that plumes, even after rising from the deep ocean floor, could remain as tightly confined as 25 km in diameter as they reached the underside of the ice. That result provides good support for the melt-through model.

While the heart of his paper on the subject supports our melt-through model, the concluding discussion twists the story to give the opposite impression. Collins, from Jim Head's group, was a coauthor. He and Goodman argue that the 25 km-wide plumes were too wide to explain the prevalence of patches of chaotic smaller than 10 km across. That conclusion is nonsense. Goodman's calculations did not address what would happen at the surface of an ice layer if a plume of warm water pressed against its base. If the plume of warm water thinned the ice enough, very small breakthroughs could easily open to the surface. Goodman's models show that melt-through is plausible; they hardly prove that small-scale breakthrough is impossible.

If Jason Goodman had presented the results of his oceanic-plume study objectively, he would have offered a useful contribution to understanding of the possible physical processes on Europa. He chose instead to do amazing contortions to get on the bandwagon.

***○ *

Theoretical geophysicists had begun modeling convection in the ice almost as soon as they realized there might be large amounts of tidal heat flowing from Europa's interior. The issue was crucial because a liquid ocean could be maintained only if the heat did not flow out too fast through the ice. The consensus is that convection is unlikely if the ice crust is less than about 20 km thick.

Beyond that conclusion, it is difficult to figure out details. Models of convection on Europa depend on making educated guesses about the unknown properties of the ice, so no matter how detailed and quantitative they are, the results will be only hypothetical.

In general, the theoretical modelers have not demonstrated the blatant biases of some of the other examples in this chapter. Because they are deeply involved in their theoretical specialty, they do however rely on others for interpretations of the images. With faith in the system they naturally accept the authoritative reports. They start from the certified fact that convection is manifested in the surface of Europa by pits, spots, and domes. None of this is true, but how can the theorists know that? Then they adjust the various unknown parameters that describe the ice layer in order to construct a model that fits the "observation." In this way, they try to infer the character of the interior.

Theoretical modeling is difficult enough, so theorists trust authoritative sources regarding the observations. The last thing that a researcher wants to hear after buying in to the establishment description, and doing all that theoretical work, is that the basic observational premise was false.

The Spanish researchers Javier Ruiz and Rosa Tejero obtained fairly typical theoretical results. They found that convection on a scale suggested by the mythical PSDs would require very specific properties for the ice. These requirements are unrealistic, so Ruiz and Tejero, in a circulated draft of their paper, wrote: "Thus, it is not possible to reconcile the heat flow results with a convective origin for the lenticulae." Furthermore, they said that "the interpretation of the lenticulae as features related to convection (Pappalardo *et al.*, 1998) is not the only explanation (Greenberg *et al.*, 1999)." As published in *Icarus*, however, this conclusion changed course by 180°, despite the fact that the results had not changed at all. The sentences quoted above were replaced by: "Thus, it is possible to reconcile the heat flow results with the lenticulae spacing predicted from a convective origin for these features ..." With this new spin, Ruiz and Tejero, too, were seated on the bandwagon.

More recently, convection modelers have begun to address the actual characteristics of pits and uplifts. Among the first was Adam Showman, a young faculty member in my department at the University of Arizona. Adam has a hard time avoiding the facts about pits and uplifts. Shortly before he joined the faculty in my department, I had taped huge color prints of Martha Leake's maps of pits and uplifts, running floor-to-ceiling, on the wall in the corridor near my office. Adam happened to move into the office exactly opposite the maps. He cannot walk out his door without getting hit in the eye by brightly colored pits and uplifts. Other than that, I have no particular influence over him. In general, Adam and his collaborators are finding so far that convection does not produce pits and uplifts like what we see on Europa.

There is no evidence that the ice on Europa is convecting, nor is there any other evidence that it is thicker than 10 km. It is time to get off the bandwagon.

The Biosphere 17

IN 1979, the simultaneous revelation of Europa's weird surface and the tidal source of its heat transformed the icy moon into a leading contender for extraterrestrial life. In the 1980s, Ray Reynolds, part of the original team that had predicted the tidal heating, led much of the research on the implications of these discoveries for the possibility of life. With his co-workers at NASA's Ames Research Center, he recognized the likelihood of a global ocean and estimated that the ice crust must be less than 10 km thick. Long before I got directly involved, he and his group had already interpreted the visible lines as cracks linking the ocean to the surface, and they had already written in detail about how that setting might support life.

That work was exciting stuff. Randy Tufts by that time had spent many years in social service work, and his stewardship of The Cave was coming to completion—largely through his efforts, it became a state park. At age forty, he was looking for a direction for the next phase of his life. Then he read about Europa's ocean and grasped its implications for life. He had an undergraduate degree in geology, he knew that the Galileo spacecraft was on its way to Jupiter, and he realized that the new images could tell whether Europa really was a contender for life. He returned to the University of Arizona, this time as a graduate student, and joined my research group as the spacecraft's data began to arrive.

The consensus of the planetary-science community was more conservative. In general, the very existence of the ocean, let alone the idea that the ice might be thin enough to create a habitable setting, was still considered unresolved. Ray Reynolds and his collaborators were highly respected, but their work on Europan life was generally considered to be intelligent speculation. So, by 1998 when a different authorized conclusion began to emerge from the Galileo imaging team—an ice shell so thick that any ocean must be isolated—the vision of Reynolds et al. seemed to be irrelevant.

Researchers who wanted to consider the possibility of life on Europa were driven to the bottom of the sea, quite literally, in a search to find plausible ways that life might be supported by the tidal heat emerging from the rocky interior. Not long before then, thriving, otherworldly ecosystems that draw their energy from the Earth's internal heat had been discovered at undersea volcanic vents on Earth. Some scientists were even suggesting that terrestrial life might have formed in those ocean-floor hot springs. So, when the scientific community was informed that Europa's ocean was isolated from solar energy, it was natural to speculate about life at hypothetical volcanic vents in Europa's sea floor.

Theorists who wanted to describe possible life in Europa's ocean met with varying degrees of success. They were stuck with two stubborn problems. First, with the ocean isolated from the oxygen at the surface, life would be limited at best. What is more, we can only guess about conditions 150 km down at the bottom of the Europan sea. There is no evidence for volcanic vents, only speculation.

But this we can say with certainty: Although we know nothing about the sea floor, we know plenty about the ice crust. Galileo images showed that the crust is like what Reynolds *et al.* had envisioned: thin enough for the ocean to be linked to the surface. As a member of my research group, Randy Tufts found himself playing a key role in developing this picture, along with myself, Greg Hoppa, Paul Geissler, and my other students and post-docs.

Taken together, all of our work interpreting Galileo data consistently pointed to the same conclusion: The icy crust has been highly permeable, and probably still is, with many types of openings at various times and places connecting the ocean directly to the surface. Cracks penetrate vertically. Heat melts open holes. Impactors poke right through. Europa's ocean is not isolated, so there are real prospects for life without resorting to speculation about the unseen seafloor.

The permeable ice layer between the ocean and the surface allows an exchange between the substances at the top of the ice and those in the ocean water. This connection prevents the difficulties that oceanic life would face if it were isolated beneath a thick, impenetrable ice crust. Life generally thrives at the interfaces between different physical conditions. The richness and diversity of tide-pool life on Earth is a good example, as is the extra density of weeds along the side of a road. Such biological communities exploit the conditions at the border, as well as the disequilibrium between the different environments.

The chemicals at the surface of Europa and in the ocean are very different from one another because they come from very different places, and even the thin crust keeps them from mixing freely. On the surface side of this boundary, molecules come apart or recombine under the bombardment of all the energetic particles and radiation within Jupiter's magnetosphere. Oxidants are continually produced by solar ultraviolet radiation and by energetic charged particles. Significant reservoirs of oxygen have been detected from the spectrum of reflected light in the form of H_2O_2, H_2SO_4, and CO_2 (hydrogen peroxide, sulfuric acid, and carbon dioxide). Moreover, a very sparse atmosphere of oxygen (O_2) and ozone (O_3) is suspended just above the surface—a fact that tells us that even more of these substances must be mixed in with the ice. (The likelihood of their presence is increased as well by the detection of O_2 and O_3 on other icy satellites such as Ganymede.) When Gaidos *et al.* argued against life on Europa, it was not because they doubted that there was plenty of oxygen, but rather that they had been told, as fact, that the ocean was isolated from it.

Organic materials and other fuels probably rain down on the surface in a continual bombardment by tiny cometary particles, as well as by the larger comets that produce observable craters. Such substances have been detected on the other icy satellites. In addition, sulfur and other materials are continually ejected from Io and transported to the Europan surface.

On the other side of the boundary, below the ice, the ocean has its own source of chemicals: substances from Europa's rocky interior, as well as those that come down through the ice. Evidence regarding the oceanic composition may be found wherever the ocean has been exposed, especially along the major ridge systems and around chaotic terrain. The Galileo color pictures (including practically all the color plates in this book) show the stuff is an orangey brown, so it is generally assumed to include some combination of sulfur compounds and organic (that is to say carbon-based) chemicals. Unfortunately, though, the visible wavelengths of light did not provide enough information to identify any specific substances.

However, Galileo's Near-Infrared Mapping Spectrometer (NIMS) did provide spectra in a wavelength range that allowed identification of some substances in the orangish brown mix, specifically sulfur-based salts in frozen brines, along with sulfuric acid and related compounds. The NIMS data could not identify specifically what additional substances were coloring the areas orangish brown, but the fact that frozen brines are part of the mix seems to confirm that it all came up together from the ocean. The ocean supplies salt, sulfur compounds, and organics to the surface, where they can mix with other substances that are already there.

At the same time, surface materials can be delivered down to the ocean in various ways. For example, consider the process of ridge formation, in which newly frozen ocean water with the various other mixed-in oceanic substances is deposited on top of the surface. In the process, former surface materials are buried, and then gradually work their way deeper as later ridges continually pile on top. The ice maintains its equilibrium thickness by melting at the bottom. Eventually, what had been at the surface is fed into the ocean, in a kind of recycling process. Occasionally, a melt-through event speeds up the delivery of chemicals from the surface into the ocean, and then resets the ice as a single layer of refrozen ocean.

For oxidants and organic fuels, rapid burial is especially important to prevent their destruction by radiation at the surface. Impacts may help with the initial burial by excavating and mixing materials at the point of impact. However, it is ridge formation that most likely provides the dominant mechanism for continually covering and protecting surface materials, and for gradually conveying all materials in the crust downward into the ocean, as strip upon strip of new material is deposited on top, along the edges of cracks.

If all this mixing happened quickly enough, the surface and ocean might have similar chemical compositions. But the distinct dark markings at the places where ocean water has most recently come up (at cracks and chaos) tell us that the ocean does have a distinct composition. So while the ice crust allows two-way transport between the surface and the ocean, it is slow enough that differences are maintained on opposite sides of this barrier.

The chemistry stays in disequilibrium thanks to production at the top and the oceanic reservoir at the bottom. In an active crack, the tidal ebb and flow of water continually transports and mixes the substances vertically. This transport runs through a huge temperature range, from a comfortable 0°C (32°F) at the base of the crust to about 170° colder only a few km above at the surface.

What we were learning about Europa restored the prospects for life in the ocean because, as Reynolds *et al.* had envisioned, the crust was proving to be permeable. We now had evidence that multiple mechanisms linked the ocean to the surface, that they all acted continually, and that all this happened recently and probably continues to this day.

The key physical processes act on a range of time scales that make the support of life a plausible idea. On a daily basis (remember a day on Europa is about 80 hours, not wildly longer than a day on Earth), warm tidal water is pumped up and down through the active cracks, thanks to the diurnal tides. Over tens of thousands of years, rotation carries each crack to a different location, where the daily cycle of tidal stress is also different. And, according to our studies of chaotic terrain, every few million years, exposure of open water by melt-through occurs at any given location.

While Gaidos *et al.* had described fundamental problems with supporting life on Europa, their analysis was based on the prevailing belief that the ocean was isolated. In fact, however, many lines of evidence were converging on a very different picture, where the ocean is linked to the surface in a set of specific ways.

Consider the profile of a crack, opening and closing with the daily tide (**Fig. 17.1**). No organisms could survive near the surface, where bombardment by energetic charged particles in the Jovian magnetosphere would

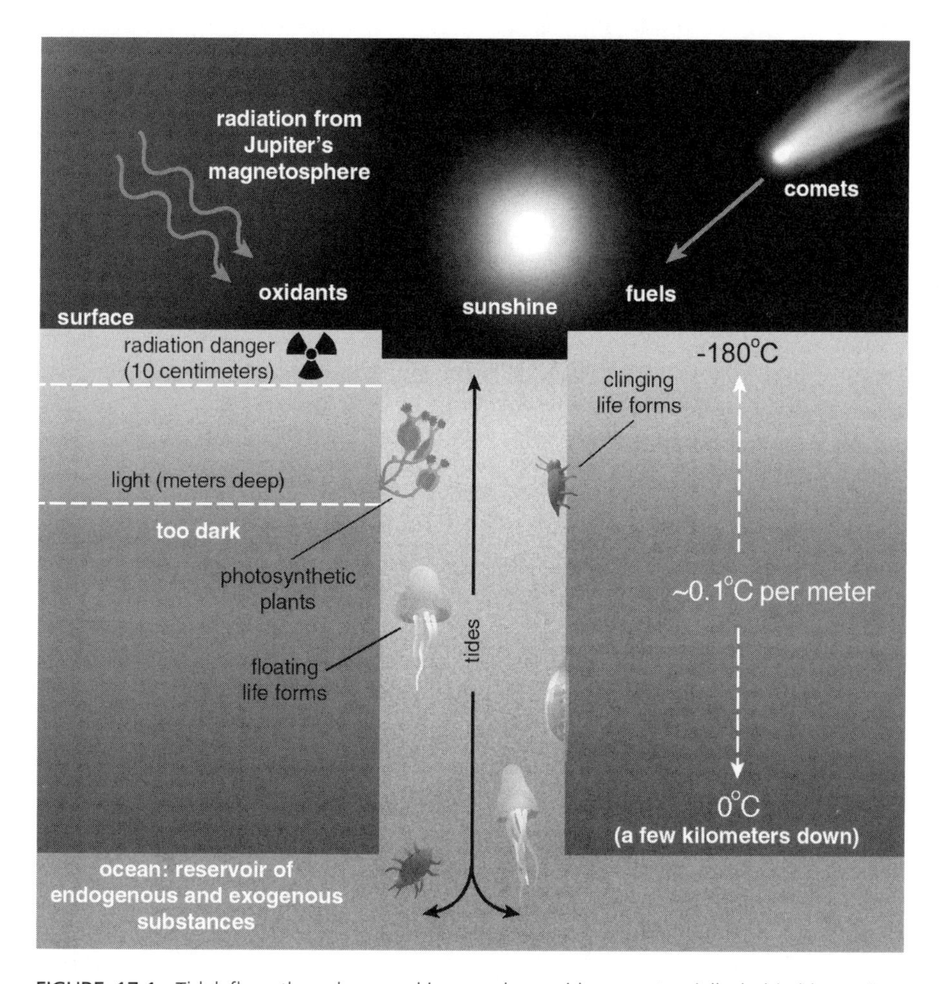

FIGURE 17.1: Tidal flow though a working crack provides a potentially habitable setting. Photosynthetic organisms (represented by the tulip icon) might anchor themselves to exploit the zone between the surface radiation danger and the deeper darkness. Other organisms (the tick icon) might hold onto the side to exploit the flow of water. Other organisms (jellyfish icon) might exploit the tides by riding with the flow. (Artwork by Barbara Aulicino/*American Scientist*, with modifications.)

disrupt organic molecules within ~1 cm of the surface. This bombardment is not very good for spacecraft electronics either. The only reason Galileo survived as long as it did was that, as it orbited around Jupiter, it spent most of its time much further out than Europa.

A few centimeters down, the ice would protect organisms from the deadly radiation. But sunlight could penetrate deeper, enough for photosynthesis as far as a few meters below the surface. For the ecosystem of the crack to exploit photosynthesis, some organisms must be deep enough to be safe from damaging radiation, but shallow enough to catch the light. They would need to anchor themselves at an appropriate depth (symbolized by the plants in **Fig. 17.1**), although they would also need to survive the part of the day when the tide drains away and temperatures drop.

Other organisms (symbolized by bugs in **Fig. 17.1**) might anchor themselves at other depths, and exploit the passing daily flow of chemicals and bits of biological stuff. As with life everywhere, their hold would be precarious. The liquid water could melt their anchorage away. Some might be plated over by newly frozen water, and embedded into the wall. Individuals that lose their grip would go with the diurnal flow to a different depth or down into the ocean. Being adapted to holding onto the walls, they might try to reattach their anchors, perhaps in a neighboring crack.

Still other organisms (symbolized by jellyfish in **Fig. 17.1**) might be adapted to ride along with the mixing flow. Many would be squeezed out of the crack each day as it is closed by tides. Some might be entrained in the slush that builds ridges along the rims of the crack, leading to probable death at the cold and irradiated surface. Luckier individuals would be squeezed into the ocean, then flow up again with the next tide.

The creatures and organic debris that are squeezed into the ocean could help feed any oceanic ecosystem. Of course, the linkage between the crust ecology and life in the ocean would go both ways, as organisms and detritus are sucked up out of the ocean into the crack during the diurnal opening phase. Oxidants, organics, and other substances from the crust, as well as the cocktail of substances from the ocean below, would be mixed and transported through the crack. The daily cycles of tidal variation and light *vs.* dark would nurture and govern the processes of life.

Life might be comfortable and routine in and below an active crack. Of course conditions would not be comfortable by human standards, but comfort is subjective. Our homes would be deadly for Europan creatures. And even on Earth, plenty of organisms can extract oxygen from water and function well at freezing temperatures—conditions in which humans cannot long survive.

A given crack is likely to remain active for thousands of years. Because rotation is nearly synchronous, the location remains fixed relative to Jupiter, and tides continue a constant daily cycle. The stable daily cycle might allow organisms to multiply and the ecosystem to prosper. Similarly, on Earth, humans have multiplied and prospered because the Earth's climate has been comfortable and stable for several thousand years. A stable and forgiving environment gives well-adapted organisms a chance to increase in number and in diversity. As long as environmental change is gradual, organisms and the ecosystem may have time to evolve and adjust.

Even where environments are relatively stable, they change over longer periods of time. We cannot count on the climate staying steady forever on Earth, and neither could creatures in a crack on Europa. Over tens of thousands of years of gradual nonsynchronous rotation, their home would move to where the tidal cycle might not work it so vigorously. The crack would seal, freezing any immobile organisms within it. Some portion of its population might be locked out of the crack, trapped in the ocean below. To survive, the organisms left in the ocean must adapt to ocean life, or move to a still-active crack. Those that become frozen into the old crack can only wait to be released.

An organism would need to hibernate for only a few million years before a melt event releases it. At any given location, chaotic terrain forms that often, on average. A few million years may seem a long time to wait in suspended animation, but bacteria have hibernated, frozen in Antarctic ice, for that long and then been revived. On Europa, frozen organisms liberated by melting would float free in the ocean and perhaps find their way into a habitable niche.

Some frozen trapped creatures might be freed in another way. During continuing nonsynchronous rotation, tidal stress would create fresh cracks through the region. They would likely cross older refrozen cracks, liberating organisms into a niche similar to their old home. That advantage is counterbalanced by the fact that release only comes at the intersections of new cracks with old ones, releasing fewer captives. They could be the founders of the ecosystem in the new crack.

In any case, the need to survive change may provide a driver for adaptation and mobility. The time scales for tidal change may help support natural selection. Those individual organisms that are able to survive the hard times and return to an active crack would then have thousands of years to reproduce in a relatively stable niche, and thus the best-adapted individuals would have an opportunity to multiply. To put it another way: Conditions in Europa's crust may be comfortable enough for life to prosper, but challenging enough to drive adaptation and evolution.

The physical conditions on Europa also provide a way for life to exist and prosper in the ocean, by providing access to necessary oxidants and linkage between oceanic and crustal ecosystems. Oceanic life would be part of the same biosphere as organisms in the crust. The biosphere might extend down into the sea floor, exploiting possible volcanic vents. But life would not necessarily depend on such uncertain resources. The crust could provide what is needed.

***O *

Then there is the question of where any life on Europa may have come from in the first place.

Even if a setting is habitable, life would need to originate in some way. The origins of life on Earth are becoming better understood, but for Europa we cannot say whether conditions were conducive to spontaneous origins. One intriguing idea may be especially relevant to a cold planet: Some theorists believe that the basic genetic material RNA requires low temperatures to form in the first place, because higher temperatures might break down the chemical structures before they grow into stable forms. If that story is correct, the possible existence of life on Europa is still more plausible. In addition to water, one commodity Europa has in abundance is the cold, so the planet is suited in that sense to originating life.

Another intriguing idea about the origin of life on Earth is that tidal cycling may have played a big role, by alternately soaking and drying the materials, helping to form the molecular building blocks of life. Europa's crust also provides tidal cycling, though further consideration is needed to determine whether they might enhance the initiation of life.

If the origins of life require sites analogous to sub-oceanic volcanic vents on Earth, there is a good chance that Europa provided those as well. Almost certainly, there has been plenty of heat generated in the interior. Volcanism at the sea floor is plausible, although we have no direct evidence for it.

Another source for the first life on Europa might have been transport from elsewhere. If some sort of "pan-spermia" seeded the early Earth, it likely would have seeded Europa as well. If life initiated independently on Earth or some other terrestrial planet closer to Jupiter, there is some small probability that colonists might have been transported to Europa, riding on rocks ejected from the planets by impacts. Some bacteria have been found to survive surprisingly well in space. These considerations increase the plausibility (perhaps from completely insane to amusingly nutty) of transport from the inner solar system. And even if theorists dismiss outright the idea of life that has already been transported to Europa, they need to consider how in the future, with human assistance, new forms of life may yet find their way there.

***○ *

Ultimately, the means by which organisms are most likely to be transported to Europa from Earth will be by hitchhiking on spacecraft. NASA was admirably cautious in terminating the Galileo mission by crashing the spacecraft into Jupiter in the autumn of 2003, rather than letting it drift around in orbit with the chance of colliding with a satellite. The habitable zone of Europa, its biosphere, extends upward from the ocean to within a few centimeters of the surface. Europa would be vulnerable to infection if living organisms plopped down on it.

When isolated ecological communities come into contact, the result can be opportunity or catastrophe, depending on your point of view. Some organisms find a new world to colonize; others are decimated or destroyed. When people are involved, access to new resources has been a benefit (at least to some of the people), while introductions of pests and disease have been obvious problems. Not only humans, but all organisms are likely to be affected. But however one views it, the changes are dramatic and they last forever.

Because extraterrestrial life may exist, planetary exploration could bring trouble if people are not careful enough. This danger was well recognized decades ago, when astronauts ventured to the Moon. When the first crews returned, they were quarantined to prevent "back contamination," the hazard that some infectious organism might have hitchhiked back with them. The safety procedures were largely symbolic. Who knew the incubation period for hypothetical extraterrestrials?

Subsequent exploration has not involved astronauts, nor have samples or hardware been returned from planetary surfaces, so back contamination has not been a continuing issue in recent missions. However, "forward contamination"—that is, the infection of alien organisms or ecosystems by terrestrial organisms hitchhiking on a spacecraft—is a distinct possibility. Consideration of forward contamination has focused on Mars since the 1960s. Now, with Europa laid open like a planetary Petri dish, the possibility of forward contamination there seems very real.

By definition, forward contamination does not affect the Earth, so why should it be of concern to us? To a large extent this question is one of ethics: Is it morally right to endanger life elsewhere? And what if we inadvertently threaten and antagonize potential enemies? That hazard seems remote, especially given that people have damaged so much life on Earth without so far having provoked conscious retaliation.

A practical concern is the more plausible prospect that an exploration campaign would contaminate a planet before completing the objective of characterizing native life there. If we destroyed or modified extraterrestrial life before we could find out about it, we would fail to achieve one of the most exciting and motivating goals of planetary exploration.

By international treaty, space-faring nations are required to appear to take precautions against contamination. In accordance with the United Nations Outer Space Treaty, the United States government retains a Planetary Protection Officer, a very nice man named John Rummel, who certainly has the best job title on our planet.

For independent assessments of policy matters related to scientific issues, the U.S. government often turns to the National Research Council. NRC is an arm of the National Academy of Sciences, a federally chartered honor society, which selects new members on the basis of whom the current members think deserves it. My former PhD advisor Irwin Shapiro once told me that what makes working in science strange is that the reward we seek is the admiration of the people we are competing against for the reward. The National Academy institutionalizes that system. Membership is perhaps the most prestigious scientific honor in the U.S.

When NASA contracts with the NRC for advice regarding planetary exploration, it gets the prestige of the National Academy, but it does not get its members. Instead, NRC puts together ad hoc committees, selected by the power brokers of the planetary science community, which produce the reports that NASA pays for. Everyone wins. NASA gets the policy guidelines it wants, with nominally gold-plated credentials, the scientists who serve on the committees for free get to influence policy and get the prestige of having been tapped by an arm of the Academy for their supposed expertise. At the same time, the Academy does a public service, while bringing in some cash.

In 2000, NASA had recognized the excitement of the discoveries by Galileo about Europa, and a campaign of missions specifically targeting the satellite was being considered. The NRC was hired to assemble a Task Group on the Forward Contamination of Europa, charged with setting the standards for protecting Europa from germs that might ride along on future spacecraft. Naturally, the party-line of thick ice was well represented. The NRC panel knew for a fact that the ice was so thick that the ocean, the only place where there might be life, was isolated from the surface.

When NRC issued its report to NASA, I could hardly have been less interested. At that point, I expected, given the players, that any authoritative panel would ignore the evidence that the ice was permeable, so the report was just one more insult. Moreover, I knew that any actual landing on Europa was decades away, when the politics driving the party line of 2000 might not be as effective. In the meantime, any concern I might have had about the short-term well-being of the imaginary creatures of Europa was addressed by NASA's actions, if not by its advisors. The Galileo spacecraft was deliberately destroyed by plunging it into Jupiter, leaving Europa safe for the present.

While I wasn't taking planetary protection policy very seriously, Randy Tufts saw things differently. He had joined the business of planetary science precisely because of his interest in life on Europa, and he came to it with an activist's temperament and experience. Nothing about Randy could let that NRC report get by. As my former student, and by then my post-doctoral associate, he pestered me to take a careful look at the report with him. Will I ever learn to stay out of trouble?

***○ *

The NRC made a specific quantitative recommendation about the maximum acceptable level of risk for contaminating Europa that NASA should adopt in planning and carrying out its exploration programs: the number used was 0.01%. As the basis of this risk percentage, the report cited a 1964 resolution of the Committee on Space Research (COSPAR) of the International Council of Scientific Unions. At Randy's insistence, we tracked the number back to its source.

In the early 1960s, Carl Sagan and Sidney Coleman derived a quantitative requirement for the sterilization of spacecraft to be used in an anticipated campaign for the exploration of Mars. The basis of the calculation was not an ethical standard with the objective of protecting Martian life, but a practical requirement meant to ensure that these missions would characterize Martian life before significantly contaminating the planet. Sagan and Coleman chose a target probability of success of 99.9%. In other words, they had no problem with destroying life on Mars, but only required that the probability of messing up Martian life should stay below 0.1% during the exploration campaign. Ethical issues aside, for the moment, this number was completely arbitrary. To meet this goal they calculated, assuming conditions relevant to Mars and making educated (early 1960s) guesses about the future exploration strategy, that any spacecraft launched to Mars had to be sterilized enough that there was less than a 0.01% chance of having any organism on board.

That evaluation was adopted by COSPAR. Although the result was given that impressive international imprimatur, it rested on a rather shaky foundation. It derived from a notion that the purpose of planetary protection is to protect the interests of scientists, rather than alien organisms. It selected the level of acceptable risk arbitrarily. It relied on pre-Space Age understanding of the planet Mars and on crude assumptions of how well a terrestrial organism might survive the trip and colonize the new setting. Finally, the prescription was based on a 1964 guess about the specifics of a future multi-mission campaign for exploring Mars.

Whether the COSPAR policy was ever appropriate for Mars is therefore highly questionable. But Randy and I were totally baffled as to why the NRC would have applied it to Europa, a completely different planet, in 2000.

And while the specifics of Sagan and Coleman's analysis are irrelevant, it did provide us a template for setting planetary-protection criteria. It demonstrates that the basis for any evaluation must be a moral or philosophical principle. Of course the standard introduced by Sagan and Coleman could be applied to Europa with appropriate calculations. But that kind of analysis of the risk is self-serving. It does not address the well-being of life on the target planet, except that it should survive long enough to satisfy human curiosity.

At the opposite extreme would be another principle: that non-interference with life on other planets would take absolute priority. Randy and I called this concept the "Prime Directive" ideal, borrowing the term from *Star Trek*, another early 1960s source. Actually, in Star Trek, the "Prime Directive" usually applied only to protecting alien societies, and even then it was readily discarded when convenient in order to advance the plot line. Yet the phrase seems appropriate here because it conveys a certain absolutism. The problem with this principle is that, if rigorously applied, it would likely bring exploration of some of the most interesting moons and planets to a halt.

Randy and I considered whether another moral principle might provide a more workable basis for developing a planetary-protection standard—one that would be more objective than the self-serving principle of Sagan and Coleman, and less constraining than the absolute isolationism of the "Prime Directive." We came up with one that we believed was objective but not absolute. It is based on the idea that there is already a process of natural cross-contamination among planets, which are thought to exchange chunks of crust from time to time after an asteroid hits and sends off ejecta into space. Living cells could conceivably survive such a journey. After all, many kinds of delicate organic molecules (including, perhaps, the very molecules that allowed life to develop here in the first place) are regularly carried to Earth within meteorites.

As long as the probability of people infecting other planets with terrestrial microbes is substantially smaller than the probability that such contamination happens naturally, exploration activities would be doing no harm. We called this concept the Natural Contamination Standard. For Europa, this standard may seem nearly as strict and confining as the Prime Directive, because natural transport of viable organisms from Earth may be so difficult that it provides an impossibly stringent criterion. On the other hand, any organisms that get on a planetary probe probably come from southern California or Florida, where most planetary spacecraft are built and launched. They are less likely to prosper in the cold, icy environments of Europa. That mundane factor may allow us to meet the stringent standard. We just have to keep any Arctic bacteria from getting on board. The actual operational requirements that follow from any underlying standard will need careful scientific study.

The National Research Council's report did not address the fundamental issues that must underlie any risk assessment, nor did it present a quantitative assessment. And it certainly did not take seriously the evidence that Europa's icy crust is permeable to the ocean below. Nevertheless, at the heart of the report was, as mentioned above, its recommended quantitative standard for planetary protection: "The probability of contaminating a Europan ocean with a viable terrestrial organism at any time in the future should be less than 10^{-4} [i.e., 0.01%] per mission." It cited the COSPAR resolution as the source of that number, which in turn had come from Sagan and Coleman. But their number had not been intended to be an overall criterion for protecting a planetary ecosystem. Instead, the number they had developed for COSPAR was meant to represent the sterilization standard for each vehicle. For Europa, the NRC applied the number to a completely differently defined probability, with no rationale. That error made it all the more ironic when their report left to NASA the responsibility for computing the sterilization requirements.

Randy and I discovered these problems in a draft of the report that had been distributed for comment. When we dutifully offered our invited comments, they were not welcomed. In fact, the NRC staff person in charge of the project was positively livid. The staffer made it clear that further consideration was impossible, because the committee had been disbanded and NASA's payment had all been spent. There was no money left to fix the report. I cannot help wondering why they asked for comments on the draft.

Randy and I published a piece on this issue in *Eos*, the news and policy journal of the American Geophysical Union. The editors included with it a response by Larry Esposito, the planetary astronomer who chaired the NRC committee. Larry is a decent person who obviously had taken on a tough job. Regarding the 0.01% figure he wrote, "the task group members reached a consensus on this value based on their collective experience and judgment. ... The best justification is that it is the result of thoughtful deliberations of the task group members." Several members of the committee privately told me the same thing, that the recommended value did not really come from the COSPAR resolution that they cited, but rather it was simply a compromise among the subjective judgments of the members of the group. The reference to the COSPAR resolution was added afterward to lend an appearance of objectivity.

From that point forward, the 0.01% standard took on a life of its own. NASA contracted with JPL to do advanced planning for future missions to Europa. As part of that planning, JPL was directed (and paid) to develop strategies for meeting the standards set by the NRC report. Engineers were assigned to determine what must be done to be sure the 0.01% limit is not exceeded.

Before NASA proceeds too much further with any Europa campaign, the scientific community needs to reopen discussions of the development of guidelines for preventing forward contamination. It needs to be based on some definable ethical or philosophical principle. The one Randy and I introduced is a candidate. Once a principle is in place, quantitative standards for mission design, construction, and operations can be developed by straightforward scientific analysis.

Another important reason for continuing discussion is that knowledge of Europa has grown since the NRC report. The evidence that Europa's icy crust has openings to the ocean is now more widely appreciated, and it has profound implications for the possibility of contamination. The Europan biosphere is far more vulnerable than if it were confined to the ocean under more than 20 km of ice.

The likelihood that the ice is thin and permeable also makes Europa an even more inviting target for exploration. Active, dynamic resurfacing could be observable, and organisms might be available for sampling at or near the surface. For the same reasons, Europa would be even more vulnerable to forward contamination than most planetary scientists might have thought a few years ago. NASA should not proceed until it has grappled with these issues in a more serious and objective way than it has in the past. It seems likely that there will be plenty of time for it.

18 Explorations to Come

During the Galileo mission, NASA decided that Europa was one of its highest-priority objectives in planetary exploration. Galileo Project Scientist Torrence Johnson consistently cited our explanation of the cycloids as the first convincing evidence for an ocean under the surface. That ocean made Europa the sexiest planet in the solar system. From a logical point of view, the thick-ice canon should have put a damper on the enthusiasm, because thick ice meant the ocean would be hard to reach and there would not be much life there anyway. But the same players continued to talk up fantasies and speculation about life in the ocean. They needed to sell the next mission to Europa.

Mars, of course, has long had its own advocates. The case for life there is equally provocative, and it has the advantage of being easier to reach, at about a tenth the distance. More recently, the hot ticket has been the Cassini mission orbiting Saturn, so the momentum shifted toward the moons Enceladus and Titan. For a while Enceladus was touted as the new Europa, because it almost certainly has some liquid water and even active geysers. But we are still too far from understanding Enceladus to know whether it is in the race. Titan, with a real atmosphere and some provocative chemistry, is a definite contender with some powerful and smart advocates.

Still, in the competition for the next flagship mission to the outer solar system, it is hard to beat the appeal of Europa with its global ocean just below the ice crust and, according to the evidence outlined in this book, connected directly to the surface. What is not widely known is that NASA actually ran a successful mission, known as the GEM, specifically targeted at the exploration of Europa. Here is the story.

As the Galileo mission was approaching the 1997 end of its operational period as funded by Congress, project management was getting nervous that further funding would be hard if not impossible to secure. With no more funding, their expensive, functioning spacecraft would be turned off. The managers' concern might seem silly. Why would Congress pull the plug on a $2-billion robot that was orbiting Jupiter and still taking pictures and gathering data? For a tiny percentage of the original investment, the project could be continued, doubling its return. Yet such wasteful mission-ending decisions had been made before. The classic example was the package of scientific instruments that had been installed on the Moon by the Apollo astronauts. It was shut down to save money in the 1970s, even though it was working fine, returning exciting and important data.

Project management at JPL and the planetary program people at NASA headquarters in Washington wanted to sell the idea of a Galileo Extended Mission to run for an additional couple of years—at least for as long as Galileo's fuel lasted and its electronics resisted deterioration in Jupiter's radiation field. There was concern that the word *Extended* would not sit well with Congress. (Generally speaking, lawmakers are not favorably disposed to pay more for projects they think they've already fully funded—preferring instead something new.) So, in a brilliant move in a silly game, the JPL and NASA managers renamed the extension the Galileo Europa Mission, capitalizing on the newfound sex appeal of Europa, suggesting the extension was a distinct mission, and preserving the cute acronym GEM. Congress bought it, we got two more years of operations, and taxpayers got more than their money's worth.

After GEM, Galileo got an additional extension, also formally defined as a distinct mission, called the Galileo Millennium Mission, because it carried the project forward from late 1999 into the new millennium, finally ending with the spacecraft's controlled, Europa-preserving crash into Jupiter in the autumn of 2003. Word games aside, GEM and GMM were really extensions of Galileo, and it is a good thing we got them. The original mission would have ended with orbit 11 in December 1997. More than three-quarters of the pictures of Europa, including most of the images in this book, were taken later.

The appeal of Europa did more than drive extensions of the Galileo mission. It also motivated NASA to develop an ambitious plan for future missions. Of course, the scientific advisory committees reported to the NASA planners the canonical version of the facts: Europa may have an ocean, but it is isolated under a thick layer of convecting ice.

The new exploration strategy was designed to address that challenge. It involved three spacecraft. The first was to orbit around Europa (in contrast to Galileo, which flew past various satellites while it orbited Jupiter) in the first decade of the new millennium, surviving about a month until its electronics were fried by Jupiter's intense radiation field. During its short life, the Europa Orbiter (as it was cleverly named) would survey the surface, measure gravity, and confirm the existence of the ocean.

In the following decade, a lander would make surface measurements in preparation for the third mission, which was planned to take place in the 2020s. This culminating mission would attack what was perceived as the crucial challenge, drilling down through the thick ice to explore the ocean below. In fact, the entire three-mission strategy was built around the assumption that thick ice was a major factor in the effort to reach the objective, the ocean with its possible life.

To help people visualize where this plan was going, JPL produced the graphic in **Fig. 18.1**. By now the JPL scientists were deeply committed to the notion of thick ice, and their artist carried it to an extreme. In the picture, the surprisingly jagged base of the ice is very close to the bottom of the ocean, so here the ice must be over 100 km thick. Volcanic vents, for which there is no observational evidence, would be the only hope for life if the ice were thick. So they are featured prominently, looking very slimy and

FIGURE 18.1: According to JPL, "This artist's rendering shows a proposed ice-penetrating cryobot and a submersible hydrobot that could be used to explore the ice-covered ocean on Jupiter's large satellite, Europa. The cryobot would melt its way through the ice cover and then deploy a hydrobot, a self-propelled underwater vehicle that would analyze the chemical composition of the ice and water in a search for signs of life." Image courtesy NASA/JPL-Caltech. (Black and white version of the color original.)

organic. The fantasy hardware includes a phallic penetrator that has released a bubbly submarine robot.

This vision has been very influential. Anticipating the perceived challenge of getting down through the ice, NASA began to fund various experimental schemes for drilling through thick ice and communicating back to the surface. Prototype robots (called "cryobots") that would drill or melt their way down through the ice were designed and tested. And an under-ice lake in Antarctica began to be hyped as a terrestrial analog to Europa's ocean.

Lake Vostok is mighty interesting itself, independent of any supposed similarity to Europa. It is huge, comparable in size to Lake Superior, but it lies below 4 km of solid ice. Its liquid water was cut off from the surface about 15 million years ago, so it may contain preserved samples of the Earth's ancient atmosphere and biosphere. So to those who bought the notion that Europa's ocean is isolated from the surface, Vostok seemed like a perfect place to develop technologies and strategies for a return mission.

Indeed JPL expected to be the prime contractor for the Europa exploration campaign, just as it had managed the Galileo and Voyager missions before. They needed the business (JPL is not part of NASA, but rather a private entity run by Caltech), so **Fig. 18.1** was really a sales pitch, packaged as education. Its message was powerful (not to mention Freudian): This set of missions will take us to one of the most exciting places imaginable, and we can overcome the challenges of getting there.

On the other hand, if you thought about it more deeply, there is little chance that we could penetrate such a thick ice lid, and if we did there would be little hope of finding life there. Because JPL (and NASA) were relying on advisors who were wedded to the idea of thick ice, they were addressing the wrong problem entirely.

Instead of trying to figure out how, in the distant future, they might drill through thick ice to reach an isolated ocean, they should have been thinking about how to find places on Europa where the ocean is at the surface, or has come up to the surface recently. Rather than worrying about drilling 20 km down, they should have been worrying about the best place to find an active crack—and, after a wait, how to pick up the frozen fish (or whatever).

With the rapid resurfacing, any Europan landing site would provide oceanic samples; the trick will be to find the freshest ones. The Europa Orbiter, should it ever become a reality, ought to have as a primary objective the location of the most likely sites for recent or current oceanic exposure. Then the lander mission could sample and examine the ocean right on the surface. Europa's biosphere could be reached a decade earlier and 20 km of ice closer than what could be expected from any expedition based on a drill-down by a "cryobot."

As it turned out, however, the issue became moot with the cancellation of the Europa Orbiter in 2002. After three years of planning, JPL finally came up with a cost estimate and shot itself in the foot. The project was unaffordable.

In order to restart plans for a return to Europa, NASA came up with a clever strategy—or a cynical one, depending on how you look at it. The space agency, in collaboration with the U.S. Department of Energy, wanted to develop a nuclear electric propulsion system. The motives are a matter of debate. The technology would certainly facilitate outer-planet exploration, removing the fuel constraints of conventional rockets, like Galileo's, as well as producing plenty of electricity for scientific instruments.

But nuclear-powered propulsion is also of considerable interest to the military. While NASA insisted that the Department of Defense had "no active interest" in the work, dubbed Project Prometheus, the Air Force worked on related technology and interacted with NASA on the development. When it was pressed by the media, NASA certainly did not rule out transferring the technology to the military.

NASA's nuclear propulsion program needed to be fronted by a presentable purpose, and its Europa program needed the money. A marriage was probably inevitable, and the offspring was the Jupiter Icy Moons Orbiter (JIMO), the flagship mission of Project Prometheus. Some people saw it as great deal for the Department of Defense, which would get the development work done at NASA's expense, but most planetary scientists were either apathetic or accepted the Faustian deal as the best they could get under the circumstances.

One clear drawback of Project Prometheus was that any return to Europa would have to wait until after this ambitious system is designed, built, tested, and perfected. The earliest return to Europa would not be for twenty years or more. Another wild card is the environmental issue. Environmentalists, at least a specialized fringe, protested the launching of the nuclear reactors on board Galileo (and the Cassini spacecraft to Saturn). Those electric generators were tiny compared with what would be launched by Project Prometheus.

Nevertheless, Project Prometheus seemed for a couple of years to be our best hope for getting back to Europa, even though some predicted the development time would be forever. By mid 2005, that proved true: Prometheus was postponed indefinitely. Like the Prometheus of mythology, the project came to a predictably bad end.

***O *

Rather than obsessing about how to reach the ocean, planners for Europa exploration should take advantage of its natural accessibility. Everything on

Europa's surface came up from the ocean not long ago. If we were to land at a random place, we could hardly go wrong. But in light of the geological processes that bring oceanic materials to the surface, some landing sites would be better than others.

When chaotic terrain forms, it replaces a section of crust with frozen ocean. The freshest chaos probably includes a representative sample of fairly recent oceanic materials, including any biological detritus or frozen organisms, if they exist. If changes in the ocean have occurred over millions of years, they would be revealed by comparison of the contents of the crust in patches of chaotic terrain of various ages, which can be determined by the degree of degradation of the terrain.

Ridges deliver samples in a different way. As they form, oceanic slush is squeezed out of cracks each day and piled over the adjacent crust. If the ocean contains biological materials, they would be laid out for inspection. I like to imagine frozen fish, but even microorganisms or a few tell-tale molecules would be revolutionary. The stuff near the top would have come up most recently. Digging down a few tens of meters into the ridge, samples would represent earlier life in the newly active crack. So just by sampling a given ridge, there might be the possibility of detecting ecological change over time.

Fresh oceanic material is also laid out at the surface as gaps open up and create the dilational bands. If a spacecraft landed on a band, samples could be taken that represent a time sequence without having to dig much at all. Organisms trapped near the edge would represent the first creatures to take advantage of the young, active crack, while those in the middle would represent the bugs that were present just before the opening was sealed shut. In this way, a record of ecological change may be preserved near the surface, ready to be sampled.

Chaos, and ridges, and dilation bands make up so much of the surface that a spacecraft landing at a random site would have a better than even chance of examining some fairly fresh oceanic material. Even if it misses these fresh terrains, it will still land on ocean stuff, the only difference being that the material has been sitting at the surface much longer. If there is life on Europa, it may be hard to miss it.

The point is that materials from the ocean are extremely accessible. Rather than obsessing about the daunting, perhaps impossible, task of drilling down tens of kilometers through solid ice, we should be considering how to exploit this accessibility at the surface. Some landing sites would be better than others. If NASA makes an effort to "land smart," choosing its landing site carefully, there may be no need to drill through the ice. The goal should be to choose a site with recent oceanic material like fresh chaos or young ridges. Instead of expending resources on technologies to drill deep down through the crust as NASA has been doing, we should be preparing to sample near the surface. And if we really land smart, we might choose a place with active oceanic exposure. A craft landed next to an active crack might only have to wait a few hours until the ocean comes oozing up.

This is not to say that it would be easy to land a craft laden with highly sensitive scientific equipment on Europa. The rugged terrain, especially in the most interesting places, would make a soft (low-impact) landing challenging. (For that reason, NASA's preferences on Mars and the Moon have always been for flat boring landing sites.) And the challenges of Europa don't end at touchdown. After surviving the landing, any conventional electronics would be fried by intense radiation. Greg Hoppa has pointed out that many of these problems could be avoided by using a penetrator, a hardened projectile designed to ram into the ice upon impact. You could land wherever you want, and the scientific instrumentation, sampling tools, analytical devices, and avionics would be buried, safe from the nasty radiation. In fact, down past the top few centimeters, there might be a chance of finding organisms or oceanic substances undamaged by the radiation. "Use a periscope to put up your antenna and to look around the surface," Greg suggests.

Despite all the talk and attention to the implausible plan to drill through the solid ice, one recent NASA-funded project recently developed technology for exploiting natural openings to get down to the Europan ocean. The legendary cave diver, explorer, and engineer Bill Stone has designed and built an autonomous underwater robot called DEPTHX that can map and navigate through odd-shaped channels. Its on-board artificial intelligence allows it to sample and analyze candidate biological material. DEPTHX has begun by exploring deep, inaccessible caves on Earth (see **Color Plate 8**), but NASA's interest in this work is with an eye toward Europa.

When will we return to Europa? After the death of Prometheus, NASA turned to an advisory committee and the research community to consider once again the scientific priorities for exploration. By the end of 2005, the response was clear: Europa remained NASA's top priority for a mission to the outer solar system. But also by then the Bush administration had declared manned visits to the Moon and Mars to be of highest importance. With that constraint, and with the government's massive budgetary deficit, there was no room for the start of a major mission to Europa in the budget that Bush sent to Congress in early 2006. For the third time, NASA's seeming commitment to a Europa mission had been withdrawn.

While NASA, as part of the executive branch of government, was carrying out the President's direction to push for the Moon and Mars, the Cassini mission was keeping the spotlight on the outer solar system, specifically on the moons of Saturn. I love Europa, but Titan and Enceladus and the rest are also quite fantastic. As usual, the scientific herd rushed to the latest thing. Strong advocates emerged for missions to Enceladus or to Titan. (To be fair several of these scientists had been talking up Titan especially for a long time. Another constituency is making a strong case for Jupiter itself, and its system as a whole, as a primary objective.) By 2007 support within NASA had strengthened for getting the planning process started again for a flagship mission to the outer solar system.

While Cassini had revitalized excitement about the outer solar system, it also produced competition for Europa. Would the next big outer-planets mission go to Europa, or would Titan or Enceladus jump in line ahead? I hope that the decision will be based on scientific grounds, but the omens are not good. As has been too common in the space-exploration program, the process has been set up as a competition rather than a scientific deliberation. First, each satellite has been assigned to a separate institution. Caltech's JPL gets Europa; Johns Hopkins' Applied Physics Lab gets Titan; and NASA's own Goddard Space Flight Center, just outside Washington, D.C., gets Enceladus. How the outer solar system got divided up this way, I'll never know. But the stakes are high. It is presumed that, when the choice is made between Europa, Titan, and Enceladus, the corresponding organization will get the contract for a multi-billion-dollar program. JPL recently hired Bob Pappalardo to lead the campaign, the role for which he was groomed ever since becoming the designated spokesman for Europa a decade earlier under the Galileo regime. If Europa is selected, JPL wins the giant contract and, it is generally assumed, Pappalardo becomes the Project Scientist.

In the spring of 2007 NASA selected four science definition teams to lay out the scientific rationale for each candidate mission (Europa, Titan, Enceladus, and the Jupiter System). NASA explicitly required that the Europa committee should not stray from the scientific premise of thick impermeable ice. So I was not surprised that Pappalardo and his mentor Ron Greeley were appointed co-chairs of the Europa committee. What was surprising was that thick-ice on Europa seemed to have been a litmus test for participation in all these committees. They were heavily loaded with scientists who were in the thick-ice clique, and no one who had challenged the party line was on any of the committees. Here is why it really matters: It is universally assumed that these committees will morph into the science teams for the selected mission.

The system rewards those who stay on the bandwagon, and blackballs those who don't.

On the other hand, as the scientific evidence has mounted that Europa's ice is thin and permeable, politically astute planetary scientists have begun to hedge their bets. I hear less often that it is revealed truth that the ice is thick and convecting. Instead, a common pundit's line is that the ice might be thick or thin, but we will not know until spacecraft return to Europa. That spin is very clever and effective. It recognizes the strong evidence that the ice is permeable, while avoiding acknowledging that the thick ice paradigm never had an observational basis. And more importantly, it sets up a scientific framework and rationale for advocating future missions to Europa.

However, the emphasis on the need for future spacecraft missions disregards the huge amount of information that we already have in hand from Galileo. In fact, more generally, there has always been a tendency for most of the attention to spacecraft data, especially images, to be paid during a short period after they are returned to Earth. Quick interpretations are made based on a small portion of the data. After that, the crowd moves on to the latest and greatest images from the next big mission. This kind of behavior is very human and understandable: Have you heard the latest? What's new? What's next?

We have seen the pitfalls of codifying the quick-look interpretations from space missions. An even more serious problem—one that is not widely understood—is that when a spacecraft returns tens of thousands or hundreds of thousands of images, most of that data is stored away and never studied in detail—or studied at all. Occasionally a graduate student or other persistent scholar might revisit old data as part of a research project, but there is relatively little funding or motivation for digging through archived material. And the older the data get, the more difficult it is to retrieve and understand. Instead, the big money and attention are lavished on the fresh new images from the latest mission. It is no wonder that the politically powerful figures in planetary science connect themselves to missions, or that their quick-draw pronouncements become scientific dogma. The mission impulse—and again, it is a very human and understandable impulse—is to declare a find or an insight or new principle, then plant a flag and move on.

But the fact is, we might not need to wait for future missions to determine the true character of Europa's crust. Already, a careful assessment of the images from Galileo has shown a picture completely different from the party line, which was based on an initial qualitative look at early pictures. Many images and features remain to be studied in detail, and quantitative studies could go far beyond what we have done so far. Emphasizing the need for future missions, likely decades in the future, to resolve these issues may be a disservice. Increased support and encouragement for detailed engagement with the data already in hand will certainly tell us a great deal more than we already know, and may well resolve the scientific issues as they have been framed by the political process.

There will certainly be plenty of time to do it right. It may be decades before a return to Europa. In the meantime, analysis of the expensive and hard-earned data already in hand will allow us to make best use of the resources for future exploration. If we confirm that we might put a lander at just the right place next to an active crack so that within a few hours fresh sea water will slosh to the surface, or that frozen sea life are spread over the surface by the variety of ways that oceanic water reaches the surface, then we would design very different mission strategies than if we believe that everything interesting is sealed off more than 20 km down.

There is no denying the need for new data. When I did, it cost me a bet. Near the end of the "Galileo nominal mission", just before the start of the GEM, I bet Greg Hoppa, Randy Tufts, and Paul Geissler that nothing much new would be discovered during the GEM. It seemed that each new image was showing more of what we already understood—the cracks, ridges, chaos, dilation, strike-slip. It was a foolish bet, because planetary exploration always turns up surprises.

Manannán crater had already proven from earlier imaging to look just like chaotic terrain, another example of a hole punched through to the ocean (as we saw in **Chapter 15**). During the GEM, a set of very-high-resolution images confirmed this similarity. But something else appeared that is unlike anything seen elsewhere on Europa (**Fig. 18.2**). A set of two dozen dark markings radiates outward from a point near the center of Manannán. They could be fine indentations in the surface or maybe just dark material. This spideroid is surrounded by cracks that form circles about 2 km in diameter around it. The Many-Legged Spider of Manannán cost me a pitcher of beer.

The spider is weird and remains unexplained and undiscussed. For me, it is a constant reminder that continued exploratiom will keep producing surprises, and at the same time we have a lot of stuff in our closet that we too easily forget.

My former student Bill Bottke works on the dynamics of asteroids, but he has followed the Europa story perceptively from a distance. He keeps reminding me that science is self-correcting. Over the long haul the truth comes out. Eventually, spacecraft missions and data analysis will provide

FIGURE 18.2: A bizarre, two-dozen-legged spider pattern at the center of Manannán crater is seen just above the center in this picture, surrounded by concentric cracks. The interior of Manannán was imaged at very high resolution (20 m/pixel) during orbit E14. This image shows an 8-km-wide central portion of the 21-km-wide crater (**Fig. 15.5**). The lumps and bumps are illuminated from the right.

enough information that the true character of Europa, whatever it is, will be undeniable.

That optimistic view is comforting, but it assumes that the institutions that have defended science during the past century will continue to do so into the indefinite future. But can we count on it? A branch of science that punishes innovation—that blackballs researchers who don't follow the party line—may lose the resiliency to self-correct. And self-correction was not fast enough for some of the individuals who made the greatest contributions to understanding the true nature of Europa. By the spring of 2000, Paul Geissler had already decided to move his focus to Io and Mars. Greg Hoppa, realizing that his job prospects in academia were grim, left planetary science soon after for a job in the aerospace industry. Randy Tufts became ill and died shortly afterwards, universally liked, but with his remarkable contributions to Europa science barely acknowledged. To attract and keep the best minds, the scientific community needs to stop penalizing new ideas, even if they go in unconventional directions. Otherwise the field itself may whither before it finishes self-correcting.

The broader scientific community may still be waiting to understand Europa, but for my small research group, the last few years of the twentieth century were the most exciting and enjoyable that we could ever imagine. Stung and disappointed as we were by the antagonistic reception, we never regretted doing what we did. The Galileo space mission was an amazing technological feat. It obtained fantastic images of Europa. Even its technical failures worked to our advantage, limiting the image sets to a size that even a small, independent group could handle. And our political marginalization ironically gave us a strong advantage. Our tiny group was free to follow the evidence where it led us, without compromise or accommodation to the social and political forces that controlled so many others. Certainly, future research and new evidence will test and pick on what we developed, refining our understanding, perhaps reversing some of our results. But we had the chance to synthesize the information from an amazing set of images within a strong, quantitative theoretical context, and to understand the workings of a planet we came to love. The real reward in science is the personal satisfaction of discovery, and we hit the jackpot. For us, we had discovered a new world.

Further Reading

Much more scientific detail can be found in my book "Europa, the Ocean Moon: Search for an Alien Biosphere"(Springer-Praxis Books, 2005). A detailed annotated bibliography can be found there as well.

To find original and up-to-date scientific papers by any of the players mentioned in this book, or by anyone else contributing to research about Europa, use the Smithsonian/NASA Astrophysics Data System (home page: http://adsabs.harvard.edu/). The ADS page for querying a specific author's publications is http://adsabs.harvard.edu/abstract_service.html.

Index

Printed in the United States of America